CacaoSource

An Emerging Sustainable Chocolate Landscape

Alain d'Aboville

in collaboration with Cherrie Lo

Typesetting by Kristina Konstantinova.

alain@cacaosource.com
cacaosource.com

First Printing, 2019
Second Edition, 2020
ISBN 978-1-7340040-2-1

CONTENTS

AUTHOR

ALAIN d'ABOVILLE

AFTER A CAREER as a management consultant and executive, mostly in Europe, in 2010 Alain started working for the US State Department in cacao producing countries such as Bolivia and the Dominican Republic. This allowed him to further his passion for chocolate and start producing his own bars. He perfected his chocolate making skills in 2014 by attending a specialized course at the University of the West Indies in Port of Spain (Trinidad). He also started giving presentations and speeches on chocolate in the US, in France, the Dominican Republic and even in Afghanistan during his last posting for the US Government.

Alain is currently working with a land owner in Puerto Rico to create a cacao plantation of high-quality beans. He also participates in building a small chocolate factory in Port au Prince (Haiti) with an existing Haitian cacao exporter. Alain has a significant audience on social media as well as online at *cacaoauthority.com*

CHERRIE LO

CHERRIE BECAME A certified chocolate taster of the International Institute of Chocolate and Cacao Tasting in 2016. Cherrie is a Hong Kong born Branding and Marketing professional in the chocolate and confectionary industry. She worked close to a decade with artisan chocolate companies, chocolate schools and international chocolate brands, such as Vero Chocolates and Pierre Hermé Paris, before moving to London to continue her chocolate journey.

Cherrie is a recognized member and a Grand Jury member and Chocolate Judge for various international competitions, such as the "Academy of Chocolate Awards", "International Chocolate Awards" and "Great Taste Awards". She repeatedly appeared on local and national TV Stations and magazines in Asia over the last decade.

Loyal to her Asian background, she hopes to introduce many major world-wide chocolate brands to Hong Kong and China. Similarly, she's looking for ways to introduce international chocolatiers to the many amazing Asian ingredients and flavors.

FOREWORD

IN THE FOLLOWING pages we draw a picture of the changing chocolate scene that will replace the current unsustainable *status quo*. The abject exploitation of entire human communities via indecent low prices, and the destruction of nature via artificial substances and methods have transformed cacao farming into an undesirable and even despised endeavor in some cacao growing regions. Unless chocolate becomes truly sustainable—from farming, to making, to packaging and sales—there may well not be a chocolate industry in fifty years. This makes change the only certainty. Among the expected adjustments, we can see higher prices to farmers, drastic reductions in the carbon footprint of chocolate making, and more educated consumers.

Undergoing deep changes in its traditional consumer base in the western hemisphere, chocolate is being tried and appreciated by a constantly growing share of the world's population. The seemingly endless increase in demand (two to five percent yearly), challenges the farmers who are already facing the many negative effects of global warming. This happens in an ever-concentrated financial world where, as detailed in a UN report, ten financial consortiums buy over two-thirds of the world cacao harvest, while two companies (Cargill and Barry Callebaut) control over half of the processing capacity[1].

Most of the cacao beans (eighty percent) produced are considered commodity beans, meaning they have low aromatic characteristics. They are traded on some international exchanges, New York being the most influential, where their price currently hovers around $2,300 per ton - close to the lowest it has ever been. The remaining beans are considered quality or fine aroma and are dealt with on separate marketplaces by brokers and directly between farmers and chocolate makers and sell between $3,000 and $4,000 per ton. Finally, a minute proportion considered specialty beans, less than ten thousand tons, sells above $5,000 per ton. This book focuses on the quality and specialty market segments because of their significant influence on the global cacao and chocolate industries.

A small number of enthusiasts worldwide—whom I call *chocolate hobbyists*—spend time and money learning about chocolate, its flavors, origins, and processes, etc. Some experiment on their own to

---

1 *intellivoire.net/savoir-les-principaux-acteurs-du-cacao-monde*

discover new processes, machines or methods that produce more and/or better chocolate sensations and emotions. In a movement reminiscent to the growth of micro-breweries twenty years ago, everywhere in the world, some of these chocolate hobbyists are becoming bean to bar chocolatiers. They excitedly attend chocolate fairs such as the *Salon du Chocolat*, in search of new flavors, new techniques, and the perfect cacao bean. Meanwhile, the large established chocolate manufacturers are slowly adopting new programs to help cacao farmers while keeping a close eye on the growing bean to bar segment.

Having met many, mostly small farmers of specialty beans in South America, the Caribbean, Africa and Asia, I was able to discern the common thread that unites them as a community, and also to identify the growing relationship they are building with those many passionate chocolate lovers all over the world.

The cacao fruit can only grow within twenty to twenty-five degrees of latitude from the Equator, but chocolate is now produced in nearly every country of the world. For historical and economic reasons, chocolate is mostly produced and consumed in the industrial world, but a growing trend of producing chocolate at the source is making small dents in the monopoly enjoyed by the ten largest firms selling chocolate products.[2]

This book is dedicated to all these new chocolate and flavor adventurers and all agriculture lovers.

---

2 *icco.org-08Feb-2018*

THE BIG PICTURE

AROMATIC HEIRLOOM CACAO PODS FROM CENTRAL COLOMBIA

A VERY BRIEF HISTORY OF CHOCOLATE

The cradle of cacao

FOLLOWING VARIOUS CONFLICTING studies starting at the end of the nineteenth century, and in particular the extensive work by Russian botanist Nikolai Valivov and the Amazonian expedition of Trinidadian agronomist F.J. Pound in the late 1930's, it is now confirmed that the cacao fruit originates from the upper reaches of the Amazon basin. Indeed, the most recent work on cacao DNA accomplished by scientists Claire Lanaud and Juan-Carlos Motomayor in the late 2000's identified specific cacao varieties in the many valleys of the tributaries of the Amazon–in an area that covers parts of Peru, Ecuador and Western Colombia.

Members of the Olmec civilization (1500 to 400 BC) were eating cacao beans as fresh fruits. In their exchanges with the Mayas and Aztecs from Central America, they brought the fruit to the isthmus region. Probably by letting the beans self-ferment, the Mayas discovered that once fermented, dried, and crushed cacao was a potent complement to their diet. Because of its relative rarity and therefore cost, and the many positive effects attributed to its consumption, consuming cacao beans quickly became a social status as well as a religious ritual. It also became a form of currency as Mayan and Aztec farmers paid their taxes to the King in cacao beans. The Aztec Emperor Montezuma was an enormous consumer of the *Xocolat* drink, claiming it provided him with immeasurable strength in the battlefield and with his many wives. The drink was a sort of cacao bean stew with chili pepper, spices, and corn flour, very different from the sweet chocolate hot milk drink consumed today.

Starting the globalization of cacao

IN 1520, CHRISTOPHER Columbus had the first European encounter with cacao while on a river in Central America (in what is now Honduras, or possibly Nicaragua). But it was not until 1585 that the first commercial cacao cargo arrived in Spain. Spanish monks, the scientists of their time, were tasked to adapt the chocolate drink recipe to Spanish taste. They discarded the chili pepper, added cinnamon, nutmeg, clover, and sugar cane. Over the next century, cacao became a popular drink in Andalusia and, gradually, in other parts of Europe.

Cacao was introduced in France by Louis XIII's

Spanish-born wife, Anne of Austria, in 1615; and a Frenchman opened the first "chocolate tavern" in London in 1650. In 1659, the French King gave the monopoly of making chocolate in France to Monsieur Caillou, who became the first European *chocolatier* on the continent. After a quarter century, his shop near the river Seine in central Paris next to the Saint Germain L'Auxerrois Church, had many competitors (it no longer exists). By then, chocolate had acquired the reputation to be an aphrodisiac, and eighteenth century French culture and art is replete with sexy and erotic references to chocolate consumption.

Louis XVI's Austrian wife, Marie-Antoinette, came to Versailles with her own chocolatier who pushed the chocolate recipe further by adding orange blossom, cream, vanilla and other aromatics. In 1776, Frenchman Mr. Doret invented a hydraulic grinder to quickly reduce large quantities of cacao beans into a paste. This drastically reduced chocolate production costs and allowed a wider distribution of the product. By the end of the eighteenth century, every European country was producing chocolate and adapting it to regional tastes.

Towards mass consumption

UNTIL 1828 CHOCOLATE remained a drink, and a rather fatty one, because of the high percentage of cacao butter in the bean. Then the Dutch chemist Coenraad Van Houten designed a process using a hydraulic press to extract the butter from the cacao paste. What was left could be blown into powder and easily melted with liquids. This was a breakthrough that allowed making "cleaner" chocolate drinks. It also made it possible to produce solid chocolate by incorporating cacao butter into the chocolate paste, alongside sugar. These new chocolate solids were initially exclusively sold to royal courts. Less than twenty years later (1847) the British company Fry and Sons offered solid chocolate to the public. The "chocolate bar" was born, starting the industrial era of chocolate. Simultaneously in Switzerland, Henri Nestlé developed a process to produce dried powered milk, which could then be added to the chocolate paste to produce the famed Swiss milk chocolate, *Chocolat au Lait*; and Rodolphe Lindt invented the modern conche, a machine for mixing chocolate sugar and dried milk to produce extremely smooth chocolate.

In 1893, American Milton Hershey discovered European chocolate making equipment at the Chicago World's Fair and quickly started producing consumer accessible milk chocolate near his former caramel manufacturing facility in Pennsylvania.

Chocolate completed its mass appeal by being distributed to soldiers on both sides of the WWI conflict in Europe. It was deemed so important that during the Second World War nearly one hundred percent of the American production of chocolate was requisitioned for the Army.

CACAO TODAY

PRODUCTION

OVER 4.6 MILLION tons of cacao beans were produced in 2018 and sold for a total of $5.2 billion, creating an $82 billion market of chocolate products. Seventy-five percent of the cacao beans are coming from Africa, seventeen percent from the Americas and the remainder from Asia. Two-thirds (sixty-six percent) come from the top three producers: Côte d'Ivoire, Ghana and Indonesia. Eighty percent of the production consists in "commodity" or "bulk" quality sold at prices established at the stock exchanges in New York or London.

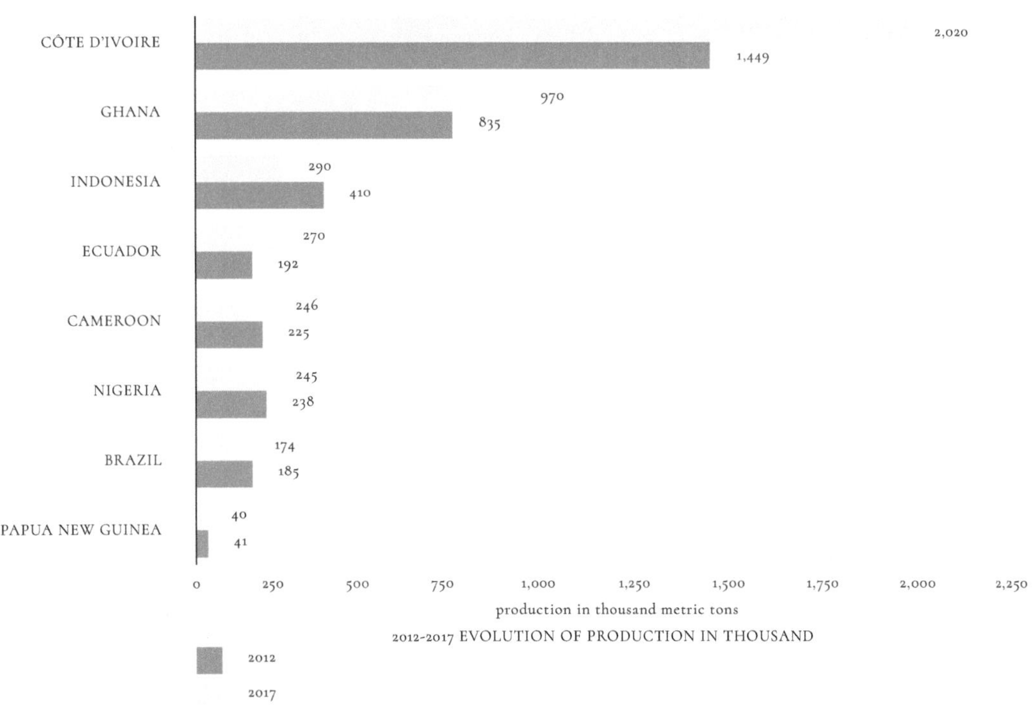

2012-2017 EVOLUTION OF PRODUCTION IN THOUSAND

RANK	COUNTRY	2015 (TONS)
1	CÔTE D'IVOIRE	1,472,574
2	GHANA	858,720
3	INDONESIA	656,817
4	CAMEROON	291,521
5	NIGERIA	236,521
6	BRAZIL	213,843
7	ECUADOR	177,551
8	PERU	107,922
9	DOMINICAN REPUBLIC	81,246
10	COLOMBIA	56,163

SOURCE: FAOSTAT, 2015

Over recent decades, cacao production has increased on average by two to three percent every year, with some ups and downs depending on weather and global economic outlook.

Because of the many threats facing cacao farming, it is not easy to forecast the availability of beans beyond five years. However, being optimistic, it is expected that the global volumes available will, on average, continue to grow well after the mid-century. Some countries, like Brazil, are predicted to make a comeback; while existing large producers, like Côte d'Ivoire, will probably plateau as newcomers appear. Some see Ethiopia, Thailand, and DR Congo among potential new "cacao nations".

CONSUMPTION

CHOCOLATE CONSUMPTION VARIES largely depending on cultural habits and consumer purchasing power. The high Swiss, German and British consumptions are based on different chocolate products—the Swiss eat mostly milk chocolate, Germans love chocolate with inclusions, and the British eat rather sweet chocolate confections.

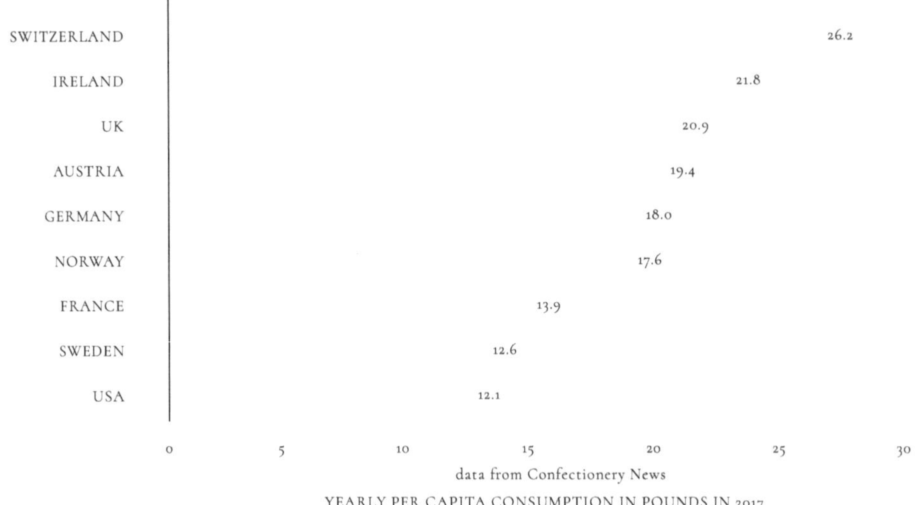

YEARLY PER CAPITA CONSUMPTION IN POUNDS IN 2017

It is worth noting that although chocolate expenditure in Africa is growing, the continent's per capita consumption is well below the existing markets. There is a link between per capita revenue and chocolate consumption.

Asia's market is expected to reach $24.4 billion by 2023, which represents a yearly increase of 7.7 percent and makes this region a tempting target for all chocolate makers, including a growing number of local Chinese chocolatiers.

Data compiled by Confectionery News, source: FoodTrending

Other cacao numbers:

- Chocolate-related sales represent $83 billion[3], 50 percent of which made in Europe.
- The top chocolate consumers per capita - UK, Switzerland, and Germany - consume about twenty-four pounds per year (nearly one kilogram per person per month!).
- Africa produces two-thirds of the cacao of the world and consumes less than 3.5 percent of the chocolate.
- Approximately five million farmers produce the 4.7 million tons of the world's harvest that become 7.5 million tons of chocolate.
- Farmers receive, on average, four percent of the value of a chocolate bar (down from sixteen percent in the 1980's).
- Asian markets represent twenty percent of the current chocolate market.

3 *https://brandongaille.com/26-incredible-chocolate-consumption-statistics/*

CACAO FARMING

THE FRUIT

THE CACAO PLANT was given the scientific name of *Theobroma Cacao* - the drink of the gods - by the Swedish botanist Carl Linnaeus in 1753 in the first recognized classification of plants. Because of the diversity of the varieties, the cacao fruit, called "pod", varies greatly in size, shape and color. However, the most common shape is a seven to ten inch ovoid (20/25 cm), similar to a rugby ball, with a hard skin showing ridges indicating when it is ripe. Inside, white pulp covers twenty to fifty individual beans connected to a central feeding rod called placenta. The pulp is rather sweet.

THE INSIDE OF **TRINITARIO POD**

Tasting it gives some indications of the aromas of the resulting chocolate because of the influence of the pulp on the beans and also because some of the flavors of the pulp are transferred to the bean during fermentation. However, the pulp does not taste chocolatey. Again, depending on the variety, the inside of each bean can be white to deep violet.

The cacao tree requires a warm and humid climate, typical of the rain forest, regular irrigation, and a neutral soil and some shade, especially in its early years. Although flowers and pods grow nearly all year round, there usually are two main harvest periods. The most important one after the rainy season and the second peak is before the rainy season stars. Cacao trees can grow up to twenty to twenty-five feet or more (height to nine meters); however, to facilitate maintenance and harvesting as well as to improve vitality, most growers usually keep trees between nine to fifteen feet (three to five meters). Depending of the variety, it takes one-half to three years for the seed to become a productive plant, and peak yield is reached in year six until year fifteen.

The typical scenario to start a cacao plantation is to select a patch of land protected from the main winds, with good irrigation and a neutral-to-slightly acidic soil (pH between six and seven). The second step consists of planting banana trees and some other tall vegetation providing shade. Soon after shading trees are in place, cacao seedlings can be planted, spaced enough to allow for growth and for future harvesting. After one year to fifteen months, the bananas can be harvested. Because after three to five years the cacao trees are strong enough, they require less shade and some of the tall trees can be removed. Satisfactory cacao yield will continue up to twenty years, at which stage it is advised to drastically prune the tree and use the rootstock to start a new tree via grafting. This will vastly reduce the time required until new production.

However, this ideal scenario is rarely applied because farmers move and change crops frequently.

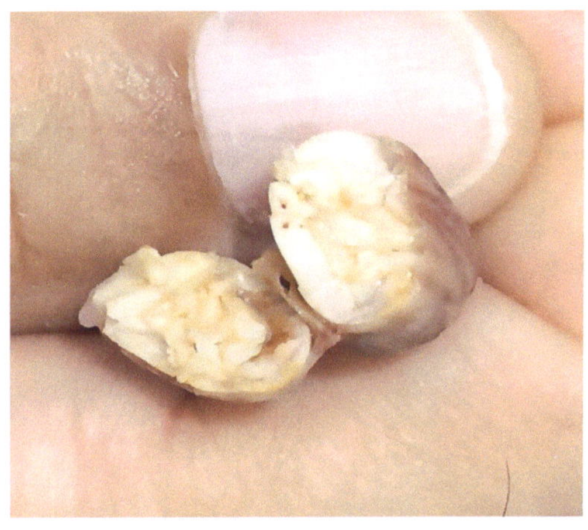

WHITE CRIOLLO **CACAO BEAN**

THREE-MONTH-OLD **SEEDLINGS**
READY FOR THE FIELD

AGROFORESTRY

IN ORDER TO fight the spread of diseases and to meet the growing demand for "natural farming" that avoids or limits deforestation, some farmers are developing mixed plantations that, beside the traditional banana and bread trees, also include diverse tropical plants such as mango, pineapple, nutmeg and sometimes even coffee. There are many advantages to this approach. As pests are usually specific to one type of plant, they do not spread well in varied plantations. As each tree produces a different set of chemicals, the soil does not become depleted of one specific nutrient needed for only one crop. Peak harvest time for each variety is unlikely to be happening at the same time, allowing farm workers to have a more balanced workload year-round, and providing a steady income rather burst of work followed by idle times.

From an economical point of view, the agroforestry method allows farmers to mitigate financial risks as it is unlikely that all the markets of each crop behave the same way. If coffee prices unexpectedly collapse, revenues from cacao or mangoes might be able to compensate.

This does not mean, by far, the extinction of mono-crop exploitations, especially in West Af-

SHADED YOUNG CACAO PODS IN THAILAND

rica or Indonesia where clusters of large plantations already exist. It is worth noting that those seemingly large plantations are often groupings of multiple small holders.

MAIN THREATS TO CACAO

AS AN AGRICULTURAL endeavor near the Equator, cacao farming is facing the usual risks of tropical farming: hurricanes or cyclones, floods, drought and pests, not to mention rodents. On top of these common threats linked to their localization and shared by everyone, the three main hazards facing cacao farmers are three fungi.

The requirement of a warm and humid environment creates an ideal ground for fungi and mushrooms of all kinds to proliferate. The most feared by cacao professionals are:

The Black Pod

DIFFERENT LOCALLY-ADAPTED VERSIONS of this *Phytophthora* kill the pods by turning them black and rotting them from the inside. If untreated, the whole tree can be damaged and die. The disease can spread swiftly as the infectious spores are carried by insects, the wind, and rain water. It is the number one threat to cacao trees worldwide as it kills up to ten percent of the trees and destroys twenty-five percent of the harvest every year.

To stop the spread of this scourge, farmers need to reduce density by planting cacao trees further apart from each other to improve air circulation and for the same reason they must maintain a clean and well pruned plantation. No stagnant water should be in or near the orchard. Rodents and other animals must be eradicated as they can disperse spores. Once infected, cacao pods, and sometimes the whole tree, must be destroyed as early as possible (burying in the ground is better than burning–which could spread spores). Some farmers use chemical fungicides, but it is a costly process that immediately removes the possibility of being "organic".

The Frosty Pod

IT SEEMS THE Frosty Pod (*Moniliophthora roreri*) has always been present in Western Colombia and East Ecuador, but it has steadily expanded to most of South American cacao fields, Mesoamerica, and is currently attacking Caribbean islands such as Jamaica and the Dominican Republic. So far, Africa and Asia have been spared. Where it hits it is absolutely devastating, reducing harvest by eighty to one hundred percent. The damaging consequences for the local communities make fighting the Frosty Pod a crucial necessity. Unfortunately, the disease first identified in 1917, starts very discreet as it grows a sort of mushroom inside the young pod. Once the gradually growing excrescence can be seen, it is already too late and the branch or the tree must be destroyed. Left unchecked, the fungus grows and breaks open the pod, leaving the beans to rot and allowing toxic spores to fly off and attack other trees. Because of its potential to virtually eradicate cacao plantations, it has become a political issue in the regions where it hits. We now know through DNA that the Frosty Pod belongs to the same family as the next cacao plague, the Witch Broom[4].

The Witch Broom

IDENTIFIED ON CACAO trees by the Portuguese in Brazil as early as the end of the eighteenth century, the *Moniliophthora perniciosa*, also called Witch Broom, is currently impacting the same areas as the Frosty Pod—parts of South America, Central America and the Caribbean. The pathogen attacks many different plants. On cacao trees it builds a sort of black and hard element on the trunk or main branches. The gradually solidifying element grows and prevents the tree sap from nourishing the branches and fruit, hence killing the tree.

Like the previously described plagues, the Witch Broom can be fought by promptly destroying the affected trees and by spreading specific chemicals. However, this approach is rarely applied because it is costly and time-consuming as the whole plantation must be spread, even the trees not currently infected. Research is ongoing to produce a hybrid seed resistant to the Witch Broom.

More cacao afflictions

THE FUNGUS ONCOBASIDIUM blocks the photosynthesis process in the leaves, which kills the tree. Many insects feed on the cacao tree, killing the leaves and eventually the whole tree. Other parasite insects like aphids and moths, among others, reduce the vitality of cacao trees by laying their eggs in the pods or sucking their sap.

CACAO PODS HIT BY THE FROSTY POD DISEASE

4 *https://link.springer.com/chapter/10.1007%2F978-3-319-24789-2_3*

Heavy Metals

OF THE HEAVY metal compounds present in food - mercury, lead, nickel, cadmium, chromium, tin and arsenic - except in specific locations, cadmium is the main concern in cacao farming.

Heavy metals naturally exist in the air, water and soil. They also come from human pollution–such as car exhaust, industrial waste, drilling and mining procedures, and of course from the use of chemical fertilizers. Cadmium levels are also increased by the "liming" process sometimes used to de-acidify soil and maintain the pH around seven. Parts of the chocolate making process also contribute to concentrate cadmium, while most of the lead comes from manipulations and packaging. As cadmium is only present in the non-fatty molecules of chocolate, the simple fact of removing the cacao butter and compacting the cacao solids to produce cacao powder concentrates the unwanted element. Heavy metals are not fully expelled by the human body which retains up to ten percent of them for twenty to thirty years; therefore, cumulative absorption is a major factor in dealing with these harmful substances. It is estimated that food and drinks provide up to ninety percent of the cadmium in our bodies (the remainder comes from the air). Heavy metals, and cadmium in particular, are linked to diseases affecting the kidneys, lungs, bones and brain, as well as some cancers. Children are most susceptible as their brain is still in a formative stage and exposure to heavy metals leads to learning disabilities. [5]

Exposure to lead has been regulated for years in many countries. But there are no federal regulations (FDA) in the US regarding cadmium. Only the State of California requires food producers to include a warning label on their products if they contain more than 4.1 mg of cadmium per daily serving. The European Union implemented a regulation for cadmium in chocolate in January 2019. It dictates the maximum levels of the heavy metal in chocolate solids and chocolate powder. The authorized amount varies depending on the percentage of cacao in the product - the higher the cacao percentage, the lower the tolerated cadmium level. This is roughly translated by cacao farmers as no more than 0.5 mg of cadmium per kilogram of beans.

Because volcanic terrains are more prone to contain cadmium, this heavy metal is present in higher concentrations in South America and the Caribbean than in West Africa. Research is on-going to develop a process to remove cadmium from cacao beans by adding specific microorganisms to the fermentation process, and by using nanotechnology techniques. However promising, these studies are not yet applicable in the field.

---

5 *https://healthfreedomidaho.org/toxic-heavy-metals-found-in-organic-chocolate*

VARIETIES

THE MYTH OF THE THREE VARIETIES

LIKE COFFEE, WHERE there are three main varieties (*Arabica*, *Robusta* and *Arabusta*), cacao has been presented as having three main categories: *Criollo*, *Forastero* and *Trinitario* .

Criollo

MEANING "INDIGENOUS" OR "local" in Spanish or Créole, *Criollo* is the native variety of Central America, where it is believed it appeared fifteen thousand years ago. The Spaniards spread this variety to other of their colonial possessions and beyond as early as the sixteen century. Besides the Meso-America region (Mexico, Honduras, Belize, etc.) and the Caribbean, it is currently also grown in Sri Lanka, Java, Madagascar, the Philippines, and Indonesia. The *Criollo* beans have a pale-white to pink color because of a low acidic content. They have a complex set of aromas with low-strength chocolatey or cacao sensations but have multiple fresh notes, from fruitiness citrus to red fruits to bergamot, earthiness and many more. This gives chocolate bars made exclusively with *Criollo* beans a long-lasting fruity after-taste. These characteristics make the *Criollo* the "King of Cacao". Despite its many aromatic qualities, this variety represents less than five percent of the world cacao harvest. This is because the *Criollo* is rather fragile, falling prey to multiple diseases that reduce its yield. *Criollo* beans are sold at a high premium and are often mixed with other varieties to create aromatic sensations in a chocolate blend.

Forastero

THIS VARIETY WAS probably named "foreign" (in Spanish) because the Spaniards saw it in South America after they had seen the *Criollo*, and as the pods looked different, they wrongly assumed it had been imported. With beans that are pink to dark violet, this variety is more acidic, which protects it from some illnesses, making it more robust and productive. As the workhorse of cacao production, it has been implanted in many countries and represents about eighty percent of the world's harvest. Because the *Forastero* name covers an extremely large number of sub-categories (*more on this later*), it has a wide range of aromatic components. The most grown sub-variety found in West Africa and Indonesia, it is a bit acidic, bitter, and can even be sour. It is well-suited to large chocolate productions using extensive machinery, processes and sweeteners.

Trinitario

THE SPANIARDS INTRODUCED cacao farming on the islands of Trinidad and Tobago as early as 1525 using the *Criollo* variety they brought from Mexico. In 1678, in order to increase productivity, they added some *Forastero* plants from Venezuela to Trinidad's gene pool. In 1727, a fast-spreading *Phytophthora* fungus infection known as "the Blast" nearly destroyed all cacao trees. After a thirty-year gap, *Forastero* seeds were imported from Venezuela. There were still some *Criollo* plants and the two varieties inter-bred naturally to produce a hybrid variety called *Trinitario*.

The *Trinitario* does have characteristics from both its progenitors. It has slightly darker beans than the *Criollo*, is definitely fruity and aromatic but can be marginally bitter. From the *Forastero*, it has gained more resistance to diseases resulting in a better yield than the *Criollo*. *Trinitarios* represent about twelve percent of the world cacao crop.

Summary of characteristics

BEAN NAME	APPEARANCE AND COLOR	# OF BEANS	BEAN COLOR
CRIOLLO	SOFT SKIN GREEN/YELLOW/RED	> 30	WHITE, OFF-WHITE TO GREY
TRINITARIO	RATHER HARD - VARIABLE	< 30	PALE WHITE TO LIGHT PURPLE
FORASTERO	HARD SKIN - GREEN TO VIOLET	< 30 > 50	VIOLET TO DEEP PURPLE

Other varieties

THIS CLASSIFICATION OF three varieties has been used for over two centuries, even as local varieties or regional names were used and hybrids were created. In 2008, biologist Dr. Juan-Carlos Motamayor led a team sponsored by the US company Mars, the US Department of Agriculture, IBM and other research centers to sequence the cacao genome. He sequenced the most widely used variety - the cultivar *Matina 1-6*, originally from Costa Rica - and nearly thirty-five thousand genes have been identified.[6]

Following this work, Professor Motamayor analyzed many other cultivars from South America, the original birthplace of cacao, and was able to identify eleven unique cultivars or varieties:

- Nacional
- Criollo
- Amelonado
- Boliviano
- Contamana aka Ucayali/Scavina
- Curaray
- Guiana
- Iquitos aka Iquitos Mixed Calabacillo (IMC)
- Marañon aka Parinari
- Nanay
- Purús

With the exception of the *Criollo*, *Nacional*, *Amelonado* and *Boliviano*, these varieties bear the name of valleys in the Upper Amazon region where they thrived without external genetic influence.

It is worth noting that because cacao flowers can easily be pollinated with pollen from flowers of other cacao varieties, natural hybridization quickly occurs in plantations and it is not uncommon to see two different varieties on the same tree. So, unless the variety is limited to one "closed" valley, it is extremely likely that, over time, it acquires genes from other cultivars.

All other cacao varieties are a combination of these eleven specific cultivars. As mentioned earlier, because of natural cross-breeding it is more and more difficult to find any "pure" cacao. Even when farmers separate the cacao trees in specific but close-by areas, after a few decades the varieties become mixed.

6 *https://www.ncbi.nlm.nih.gov/pmc/articles/PMC4053823/*

FINE AROMA VS COMMODITY

80 PERCENT OF THE BEANS HARVESTED ARE NOT CLASSIFIED AS "FINE AROMA" CACAO, THEY ARE "BULK" OR "COMMODITY"

COUNTRIES	PERCENT OF FINE FLAVOR
BELIZE	50%
BOLIVIA	100%
COLOMBIA	95%
COSTA RICA	100%
DOMINICA	100%
DOMINICAN REPUBLIC	40%
ECUADOR	75%
GRENADA	100%
GUATEMALA	50%
HONDURAS	50%
INDONESIA	1%
JAMAICA	95%
MADAGASCAR	100%
MEXICO	100%
NICARAGUA	100%
PANAMA	50%
PAPUA NEW GUINEA	90%
PERU	75%
SAINT LUCIA	100%
SÃO TOMÉ & PRINCIPÉ	35%
TRINIDAD AND TOBAGO	100%
VENEZUELA	100%
VIETNAM	40%

ICCO "FINE AROMA" COUNTRIES 2015

WHAT ARE "FINE AROMA" BEANS?

THE INTERNATIONAL CACAO Organisation (ICCO)[7] has a panel focusing on "Fine Aroma" cacao. Although the panel meets yearly, its latest report including a list of the fine cacao producing countries dates from 2015. At that time, the panel identified twenty-three countries, out the approximately forty cacao producing ones, that produce "Aromatic" or "Fine Aroma" beans. The list includes an estimate of the percentage of fine aroma crop in each country.[8]

Note that none of these countries are large producers. The list also ignores countries that are not ICCO members (like the Philippines).

Although these percentages have changed somewhat, the overall ranking remains true, except for Ecuador which has drastically increased its production of the high yield variety CCN-51, (now about fifty percent of the crop) which does not qualify as "Fine Aroma".

---

7 *https://www.icco.org/*

8 *https://www.icco.org/about-cocoa/fine-or-flavour-cocoa.html*

WHAT IS "FINE AROMA"?

INITIALLY, THE VARIETIES of *Criollo* and *Trinitario* readily qualified as "Fine Aroma", while the *Forastero* did not. However, there is more than variety to make a "Fine Aroma". The actual flavor trumps the genetics. For example, the *Nacional* and the *Arriba* varieties grown in Ecuador both belong to the *Forastero* category but are recognized as "Fine Aroma"; whereas some *Trinitario* grown in West Africa are not. As the flavor of cacao results from more factors than its genetics alone (*more on this later*) there is a subjective judgement made by the ICCO panel, taking into consideration the taking into consideration the whole "terroir" identity, the off-flavors and other aspects.

WHAT ARE THE "COMMODITY" BEANS?

STRICTLY SPEAKING, "COMMODITY beans" are the cacao beans quoted and sold on exchange platforms like in London and New York (managed by the New York Board of Trade). In effect, the price set at these exchanges applies or is referenced on most of the cacao deals of significant amounts. As these transactions are sealed without actually looking at the product, it is assumed that the cacao beans are "standard" in conditions (level of humidity, percentage of unfermented beans, rotten beans, etc...), and in quality (aroma, fat content, etc.). This means their quality, or lack of it, has no impact on the price. The quotations system leads to prices for future deals based on expected price levels over time, turning cacao into another financial product. This also incentivizes large buyers and brokers to stock or sell cacao beans in an attempt to obtain a better price. Although some cacao professionals believe these speculative actions have contributed to the extremely low price of bulk cacao since 2016, I doubt it. Despite the gigantic size of some cacao warehouses in Holland and Belgium and elsewhere, they can only hold a very small fraction of the 3.5 million tons of beans that make up the bulk market, limiting greatly their capacity to move the market price one way or another.

From the farmer's point of view, the "commodity" beans are all those that have no qualitative characteristics that they can use to get a better price. Unfortunately, more often than not, the buyer is looking for "Commodity" beans and the farmer accepts the price despite having some aromatic varieties.

Depending on the production region there can be various levels of quality. For example, the main transformer of cacao beans in the Dominican Republic, Conacado, sells its beans at five distinct levels of quality and price:

- *Bio Swiss* - Homogenous aromatic variety of organic beans grown in agroforestry farms with a high fermentation percentage and careful treatment.
- *Biodinamico* - Agroforestry and organically grown beans of varied types, with a thorough post-harvest treatment.
- *Organico* - Varied beans, no chemicals, standard fermentation.
- *Convencional* - Any beans including chemically grown ones, standard fermentation and drying procedures.

CONACADO QUALITY LEVELS

Everything else, including unfermented beans, and sometimes deficient beans, is called *Sanchez* and are sold at an even lower price.

Sanchez is the name of the small port on the Samaná Bay, which used to be the main loading harbor for cacao from the northeast of Hispaniola. It was equipped with wooden jetties supporting a railway system carrying the cacao bags all the way to the ships. The method of shipping in containers, the opening of a large container harbor near Santo Domingo together with road improvements have rendered this infrastructure useless, but its remains are still visible.

HYBRIDS AND GEOGRAPHIC ORIGINS

ALTHOUGH DNA SEQUENCING is key in determining varieties, it has not yet been connected with every possible organoleptic characteristic. This work is being undertaken at various universities and research centers. As there are other factors influencing the flavors of the cacao bean it is hard to correlate any specific gene with a specific aromatic characteristic.

Hybrids

USING A CLASSIC pollination method, agronomists have been creating hybrid beans for over a century. Until recently, their goal has always been to increase the yield of cacao trees by reinforcing their resistance, speeding growth, or increasing other characteristics such as the fat content or the overall size of the fruit - a straight forward "more is better" approach, with no concern for the aromas. The relatively recent craze for "aromatic chocolate" has raised consumer's awareness toward flavors, and some of the current research now includes organoleptic markers.

A limited number of research centers have produced aromatic hybrids recognized worldwide. For example, the British Imperial College, working in association with Latin American universities like the Trinidad-based University of the West Indies (UWI), created a family of aromatic beans. This is the case of the famed *Imperial College Selection 95 (ICS-95).*[9] This hybrid has a high *Criollo* content and is therefore fruity but remains rather sturdy and has a good yield.

The *Centro Agronomico Tropical de Investigación y Enseñanza (CATIE)* in Costa Rica is another renowned creator of aromatic hybrid beans. Chocolate bars made with its C-4 beans regularly win international prizes and top rankings at the *Salon du Chocolat* in Paris and elsewhere.

Many other research centers such as the French Agricultural Research Centre for International Development (CIRAD), and the United States Department of Agriculture (USDA) provide hybrid seeds and maintain cacao germplasm centers.

9 *https://www.c-spot.com/atlas/chocolate-strains/cultivar-strains/ics/*

Geographic origins

ANOTHER FACTOR INFLUENCING cacao aromas is their origin or their *terroir*. Similar to wine or grapes that develop distinct properties depending on the field they grow in, cacao trees are influenced by the nutrients they absorb from the soil, the surrounding plants, the quality of the water irrigating them, etc. This whole environment is also called the *terroir*. Although it is clear that the same variety has different tastes when it comes from different locations, the underlying chemical reasons are not fully known. Regarding aromas, studies are ongoing and it is suspected that the amount of zinc in the soil has the highest consequence on the fruitiness aromas.

As cacao trees do inter-pollinate easily and therefore are increasingly cross-bred, it is becoming more important to differentiate cacao by its origin rather than by its supposed variety which can sometimes be regarded as a "work in progress".

GMO

THERE CURRENTLY ARE no Genetically Modified (GMO) cacao seeds available. Mark Guiltinan and Siela Maximova from the University of Pennsylvania are part of a team including many research centers and companies working on a GMO seed. Using the gene-editing tool CRISPR developed at UC Berkeley, by biochemist Jennifer Doudna, and the Cas9 protein, the team has been able to delete the TcNPR3 gene which diminishes the capacity of the plant to fight diseases. Other genes have been modified to bring flavors, increase growth, etc. So far the team has produced over one hundred million clones that are being tested in Indonesia. At this point, the team is satisfied with the way the clones behave but more study is still needed before a seed can be exploited–in particular on the way cacao trees mature, on the characteristics of their seeds. etc. The pressure is on to design a seed that could resist the plagues that are now spreading faster because of global warming, while being highly productive and remaining aromatic.

At this stage I have not heard of attempts of adding external genes or substances to the seed, but to *only* annihilate or amplify the effect of existing genes in order to obtain desired outcomes. If this remains the case, the modifications would be similar to multiple hybridizations and therefore not necessarily bad for human health.

However, my uneasiness about GMO in general stems from the very concept of patenting living organisms, which generates a cascade of issues, problems and societal changes. A major one is changing the social relationship between farmers and seed providers, turning low income farmers who can hardly live decently, into financial dependents of large international corporations. Another important consequence is that by promoting one variety over others, the seed pool (or diversity) would gradually diminish.

THE CASE OF THE CCN-51

THE HYBRID CCN-51 is a unique case because of its tremendous yield, low organoleptic characteristics, and the constant controversy it brings.

In the early 1960's Ecuador was facing a particularly damaging infestation of the mushroom *Ceratocystis cacaofunesta* (locally called in Spanish *Mal de machete*–because it is thought to be spreading through the machete used to harvest the pods). This fungus, which attacks many other plants like potatoes and pomegranate, is deemed indigenous to the north of South America, Central America and the Caribbean. Suspecting El Niño was the culprit of this aggressive malignant wave, agronomists believed the menace would return regularly, and started working on a hybrid cacao which would be resistant to that strain of disease. From a plantation in Naranjal, south of Guayaquil, the Ecuadorian botanist, Homer Castro, used among others, the fruity varieties *Iquitos IMC 67* and the *Imperial College 95*, both regarded as aromatic varieties. He later included other varieties he'd discovered in eastern Ecuador, and it is at the fifty-first attempt that the hybrid seed displayed the resistance and productivity characteristics he was looking for. The *Colección Castro Naranjal - 51* was born. This hybrid seed was first used on plantations in the mid-1960's in Ecuador. It was quickly adopted by Ecuadorian farmers instead of their ancestral *Nacional* because it is extremely productive, yielding up to 4 tons per hectares; and it is free, because Mr. Castro did not patent his work before unexpectedly dying in a car accident the late 1980's.

Thanks to the sequencing of the cacao genome we now know that the Hybrid CCN-51 is made of IMC (45.4%), Criollo (22.2%), Amelonado (21.5%), Contamana (3.9%), Purús (2.5%), Marañon (2.1%), and Nacional (1.1%). [10]

The CCN-51 controversy

THE DISCORD ON the CCN-51 mostly stems from its weak organoleptic components. Judgements of the *CCN-51* flavor range from "At best has no taste; when it has some, it is bad", to "Not bad, but it's not interesting", to "Promising when properly processed". The fact is that no chocolate maker boasts using the *CCN-51* and no chocolate bar containing the hybrid has won a prize in the many chocolate competitions that now exist. It also does not qualify as *Fine Aroma* in the ICCO classification.[11]

As a responsible bean to bar chocolate maker myself, I have other objections to farming the *CCN-51*. As it does produce up to five times more and larger pods than the ancestral aromatic *Nacional* (in Ecuador), it sucks up to five times more nutrients from the soil. Similarly, it requires a greater amount of water. So it is virtually impossible to have a large *CCN-51* field without irrigation, and without using extra nutrients - organic or chemical. These are extra expenses for the farmers who are already among the poorest people in their region. Because of its flavor, or lack of it, the *CCN-51* commands a reduced price than aromatic cacao, requiring farmers to work harder to obtain the same revenue. This means cutting more, carrying more weight on their back, moving more tools, etc. Finally, the *CCN-51's* natural expansion and swift cross-breeding is turning the cacao regions where it grows into mono-variety plantations. Less than half of Ecuador's crop is now considered *Fine Aroma* (compared to seventy-five percent in 2015).

Regarding aromas, the proponents of the productive hybrid claim that good flavors can be developed with a specific post-harvest treatment–taking into account the important amount of mucilage (the white pulp) around the beans and the larger beans themselves. It is true that some attempts with longer fermentation time and slow-drying methods have result-

10 *https://toakchocolate.com/blogs/news/what-is-the-best-cacao-variety-in-the-world*

11 *https://www.icco.org/about-cocoa/fine-or-flavor-cocoa.html*

ed in improved aromas. They also argue that eighty percent of the cacao is not *Fine Aroma*, and therefore there is a need for other types of products. And their supposedly definitive argument is that the advent of climate change requires making the best use of every square foot of green land, and the higher productivity of the *CCN-51* meets that requirement.

All this may be true. By definition not everything can be *Premium*. However, I am convinced that the *CCN-51* does not provide what local farmers desperately need: a better life via increased revenue, not increased work and financial risk to maintain their meager revenues.

PINK CHOCOLATE

ON SEPTEMBER 5, 2017, the largest cacao bean processor in the world, the Belgo-French conglomerate headquartered in Switzerland, Barry Callebaut, launched a pink-colored chocolate in Shanghai. The company called it "Ruby chocolate" and stated it will soon become the fourth type of chocolate after dark, milk and white chocolate. Although rather secretive about its creation, Callebaut revealed it was not a new cacao bean variety, nor did it contain any GMO material. The company announced it was the result of a research program started in Belgium and in France as far back as 2004. Insisting Ruby chocolate was all natural, without any berry flavoring or added coloring, Callebaut has not detailed the special process it has patented to produce it. On its website, the company states that, "Ruby chocolate contains a minimum of 47.3 percent cacao solids calculated based on the Codex Chocolate Standard. On the total product Ruby chocolate contains a minimum of 32.5 percent cacao solids".[12] This means two-thirds of the Ruby chocolate is not coming from cacao beans.

RUBY CACAO WAFERS FORM TRADER JOES 2019 VALENTINE'S DAY

Nestlé was the first to use the Ruby chocolate in its Hershey's Kit Kat® Bars sold in Japan, where they already have a selection of flavored Kit Kat. The London-based Fortnum & Mason food store started selling a Ruby chocolate bar in 2018, and more makers are entering the field. In early 2019, Callebaut started selling to American chocolate makers. However, as it stands, it cannot be called "chocolate" until agreed by the FDA, and the component responsible for Ruby's color and high acidic content is

12 *https://www.callebaut.com/en-OC/chocolate-cocoa-nuts/chr-r35rb1/ruby*

not currently accepted in chocolate. This might be why the American food retailer Trader Joe's called its 2019 Valentine's Day special Ruby confection "Cacao Wafers"[13] and not "chocolate wafers". Early reviews of this product characterize it as "fruity and reminiscent of white chocolate, but also has some raisin-like notes". It is worth noting that until now all the Ruby bars have a cacao content of less than forty-five percent. Most people who have tasted Ruby, including me, find it very similar to a sweet raspberry-flavored white chocolate bar, without any of the classic chocolatey sensations.

The lack of information from Callebaut about the origin and process of its Ruby product has led to many suppositions. Because in 2009 Callebaut registered a "cacao-derived material" patent using unfermented cacao beans, some chocolate specialists believe Ruby is made of unfermented beans, probably *CCN-51* (which has a natural pinkish color). Callebaut stated the beans for its Ruby product come from the Côte d'Ivoire and Brazil, where the *CCN-51* is not prevalent, and also from Ecuador, which grows most of the *CCN-51* currently available.

As a matter of principle, I like to know what I put into my body and resent the lack of information about Ruby's origin or process. Additionally, I find it far too sweet and not enough chocolatey for my taste.

As I write this chapter, in June 2019, Barry Callebaut just announced it applied and recently obtained a "temporary marketing permit" in the US for its Ruby cacao. This means the company is "temporarily" authorized to market and describe its Ruby cacao product as "Ruby chocolate". I believe that by writing in their promotional brochure "Made from Ruby cacao bean", the company is misleading consumers because it wrongly implies that there is such a thing as a bean variety called "Ruby". Further down its leaflet, Callebaut claims that, "Ruby is made from bean to bar without added colors or flavors". Semantically true, this statement will further disorientate buyers by making them think that Ruby is a bean to bar product similar to craft chocolate, which it is not. The rest of the information provided concurs with the suggestions that the Ruby cacao is the result of a post-harvest process fermentation.

---

13 *https://www.traderjoes.com/digin/post/ruby-cacao-wafers*

CERTIFICATIONS

CERTIFICATION IS LIKE an insurance contract in which a risk is transferred from the buyer of the beans to the certification structure. The cost of the contract is usually spread among various participants. It does not, as such, guarantee a reality in the field, such an increase of farmer's incomes. After over twenty years or more of efforts and communication by chocolate companies, cacao producing countries and noisy customer complaints, a recent study commissioned by Fairtrade International determined that fifty-eight percent of Fairtrade certified cacao farmers in Côte d'Ivoire lived with less than \$1.9 per day, which is the definition of extreme poverty. Furthermore, over three-quarters (78%) of Ivorian cacao farmers do not meet the local living income level.

The main certifications used in cacao and chocolate are:

THE FAIRTRADE LABELLING ORGANIZATION (FLO)

THIS ORGANIZATION WAS created in 1997 in the UK, as a standard-setting organization. As presented on its website[14], Fairtrade's mission is to set standards in order to improve the lives of all the workers involved in producing all kinds of agricultural products worldwide. To that end, the organization certifies companies and businesses meeting the standards it has defined. In this pursuit, Fairtrade works with companies, lobbies governments, works with farmers, and drives awareness with the public. Widely recognized and used in all kind of industries, Fairtrade's work covers over 1.66 million farmers and workers in 1,411 producers' organizations (farms and coops). Where it operates, the organization works with local officials and representative bodies to set minimum prices for the goods, and to identify improvement projects. Additionally, FLO promotes and participates financially in projects identified in coordination with the local communities, such as building schools, improving roads, implanting sewage systems, etc.

Concerning cacao, Fairtrade is currently operating in Belize, Bolivia, CostaRica, the Dominican Republic, Ecuador, Haiti, Nicaragua, Panama, Peru, Cameroon, Ghana and Côte d'Ivoire.

14 *http://www.Fairtrade.org.uk*

THE MAX HAVELAAR FOUNDATION

INITIATED IN 1993, the Max Havelaar Foundation[15] is a Dutch organization working to guarantee small farmers in developing countries a fair price for their produce. The Foundation applies the Fairtrade standards in the Netherlands, and performs national campaigns to raise awareness of the role of Fairtrade to businesses as well as consumers. It is one of the nineteen Fairtrade national organizations covering twenty-four countries.

UTZ AND THE RAIN FOREST ALLIANCE

PREVIOUSLY TWO SEPARATE organizations working towards similar goals, UTZ and the Rain Forest Alliance have recently merged (2018),[16] "This merger was in response to critical challenges facing humanity, deforestation, climate change, systemic poverty, and social inequity," according to their website[17]. Programs are at the local and individual farmer's level, focusing on climate change mitigation via sustainability, on traceability, and gender discrimination. The new organization is active in farming coffee, cacao, tea, and hazelnuts, and works with many major brands such as IKEA, Mars, McDonalds and Nestlé, and many others like the Swiss retailer Migros, the British coffee roaster Taylors, and many more.

THE IFOAM - ORGANIC

THE IFOAM WAS founded in Versailles, France in 1972 as a tool for the then many unrelated organic agriculture associations to coordinate their actions and create a real momentum toward organic farming. Now headquartered in Bonn, Germany, it groups one hundred twenty national affiliates to promote organic farming worldwide. The general assembly meets every three years and elects a world board responsible for the coordination and implementation of IFOAM projects. Besides setting standards for organic recognition and certification, the organization runs many trainings and awareness programs focused on practitioners on the ground.

SUSTAINABILITY

SUSTAINABLE FARMING, BE it for cacao or any other plant, consists of applying farming methods that are economically viable, socially inclusive, and ecologically durable. The current quest for sustainable agriculture is the result of decades and centuries of complete disregard for the long-term consequences of the over-productive farming practices promoted, and sometimes imposed, by various interest groups all over the world and all along the supply chain.

15 *http://maxhavelaar.nl/english*

16 *https://utz.org/who-we-are/*

17 *https://utz.org/who-we-are/about-utz/*

Now that the limits of the earth's resources are becoming visible and local governments and international associations have become more critical, most large cacao companies are turning to actual and real sustainability projects. But due to old and deeply-rooted societal behaviors at all levels and because of the skepticism of a small percentage of the people involved, the needed changes have been very gradual and slow to implement until now.

For example, all the large chocolate makers have started some kind of sustainable programs twenty to thirty years ago (or more). However, child labor remains high in West Africa cacao plantations, deforestation is a huge issue in Indonesia, Brazil and other countries, and the average age of cacao farmers keeps on increasing because, in part, the share of chocolate revenues left to farmers has never been lower than it is now.

Acknowledging this diagnostic, in September 2018, one of the largest cacao buyers and chocolate makers, Mars, launched yet another program called "Cacao for Generations", pledging to spend $1 Billion over the next decade – equivalent to $280,000 every day. To further prove its commitment, the company adds this undertaking to its existing project "Sustainable in a Generation Plan" started in 2017.[18] The program's goals call for Mars to buy one hundred percent of its beans from "Responsible Cacao Program" by 2025, to increase the premium price paid to farmers, to eradicate child labor, stop deforestation, and implement other positive changes beneficial to farmers.

Mondelez's UK brand Cadbury has also announced ambitious sustainability programs in Nigeria, and elsewhere, that include farmer's revenue goals.

Although a look at the results from previous similar announcements and pledges from cacao industry leaders, as well as local politicians, could justify cynical congratulations, I believe that this time we are going to see real changes on the ground. Why this time? Because the "heat of the fire" is now being felt strongly enough by everyone, for the major actors to pursue tangible, on the ground results rather than just feel-good speeches.

Sustainability and organic

SUSTAINABLE AND ORGANIC are indeed different things. "Organic farming" has a recognized definition that leads to an official USDA certification in the US and also in the European Union. Although the EU Certification program covers a wider scope than the American one, they are both exclusively focused on avoiding any non-natural substances in the farming process. However, they are not looking at the global picture, meaning that the organic certification has no re-

18 *https://www.prnewswire.com/news-releases/mars-launches-new-cocoa-sustainability-strategy-300714181.html*

quirements regarding the well-being of farmers and animals, is not concerned about community involvement, water and fossil fuels consumption, greenhouse emissions, or packaging. Therefore, a plant or fruit can easily qualify as organic while not being sustainable.

The ISO Standards for sustainable and traceable cacao

IN MAY 2019, the International Standard Organization, based in Geneva, published the ISO standard 34101 series [19] with the goal to professionalize cacao farming in order to improve farmers' livelihoods and working conditions. As detailed on their website, the development of the ISO 34101 series involved all the stakeholders in cacao farming and chocolate production. Other standardization and technical organizations collaborated with the ISO committee, such as the European Committee for Standardization and ISO's members for Ghana and the Netherlands.

This welcomed development will certainly help in adopting truly sustainable practices in cacao farming. Like all ISO certifications it requires a process which will, no doubt, incur a cost for farmers. However, in every sector the ISO approach is extremely thorough and comprehensive. This development is likely to create trust among chocolate professionals and consumers. We hope many of the large companies currently self-certifying their cacao will adopt this formal standard which has a universal meaning.

Voluntary Sustainability Standard (VSS)

UNTIL MAY 2019, sustainable cacao farming did not have a formal standard definition and could therefore be seen as a school of thought or a way of life. Unlike "organic", "sustainability" is concerned with the overall impact of agriculture, well beyond just the food product. In the absence of one universally-recognized standard and because stakeholders all along the supply chain were interested and seeking "sustainable" cacao, many standards have been created by manufacturers and retailers; but until May 2019, there was no recognized international organization or body structure enforcing a universal standard. This situation pushed large companies to self-certify their products as sustainable, and led to confusion on the significance and usefulness of applying and using sustainable practices. The Voluntary Sustainability Standard organization[20] was, and continues to be, working on standardizing and promoting sustainability processes—and is showing a lot of progress. In practical terms, to be "sustainable" tends to require small operating teams applying socially aware and eco-friendly policies.

---

19 *https://www.iso.org/news/ref2387.html*

20 *https://www.iisd.org/topic/voluntary-sustainability-standards*

THE CHOCOLATE PLAYERS

FROM FARMERS TO CONSUMERS, THERE ARE A MULTITUDE OF PEOPLE INVOLVED IN THE CREATION AND CONSUMPTION OF CHOCOLATE.

THE GROWERS

OUT OF THE 1.5 billion farmers in the world, it is estimated that four and half to five million of them grow cacao beans, deriving some or all of their income from this activity. About ninety percent of them are smallholders with less than twelve acres (five hectares). Together they provide work - and revenue - to about fourteen million people[21], making cacao the main livelihood for forty to fifty million individuals once their family and close community are included.[22]

Four West African countries (Côte d'Ivoire, Ghana, Cameroon and Nigeria), produce seventy percent of the world's harvest. In that region, the average age of cacao farmers is currently fifty-one years old; whereas life expectancy is barely above sixty. Less than fifty percent of the inhabitants are literate because there are very few learning opportunities. South American cacao farmers fare only a little better.

According to the ICCO the average cacao yield is three hundred and fifty kilograms per hectares. However this covers wide differences, from two hundred kilograms in Ecuador to fifteen hundred kilograms in Sulawesi, Indonesia. It is around four hundred kilograms per hectares in West Africa. Productivity depends on many factors with variety, age of the trees, and quality of the pruning work being the most important ones. Varying upon location and treatment, up to thirty-five percent of the harvest is lost to diseases and pests.

The twenty percent of the world crop which qualifies as Fine Aroma is mostly produced in South America, the Caribbean, and parts of Asia.

With just a few exceptions – Hawaii, Northern Australia, etc. – cacao growers are in low-to-intermediate income countries and are usually living in remote areas with limited infrastructure.

---

21 *https://www.icco.org/faq/57-cocoa-production/123-how-many-smallholders-are-there-worldwide-producing-cocoa-what-proportion-of-cocoa-worldwide-is-produced-by-smallholders.html*

22 *http://www.worldcocoafoundation.org/wp-content/uploads/files_mf/14855360702016ResearchSymposium_Day3IITA.pdf*

THE "NEW" FARMERS

AS THE CURRENT generation of cacao growers is aging, a new type of cacao farmer is slowly emerging. This new generation focuses on quality through technology to improve revenue. Usually more educated than their parents, the young or newcomers to cacao farming tend to regard farming as a social responsibility rather than just a revenue-generating activity. As the whole market is geared towards volume instead of quality, they are facing a lot of challenges—at local level, with government entities, and with buyers. For example, challenged by their farmers-voters, many small producing countries have dismantled their National Cacao Board and/or have removed the export monopoly they once enjoyed. Because quality requires growing varied and aromatic varieties, these new developments are more prevalent in South America, the Caribbean, and some places in Asia, than in Africa and Indonesia. In those regions, history, local culture, and the pressure from the large chocolate industrial buyers have created a totally different situation.

COUNTRY / REGION	WORKERS (MILLIONS)
WORLD	**14.00**
AFRICA	**10.50**
CAMEROON	1.60
COTÉ D'IVOIRE	3.60
GHANA	3.20
NIGERIA	1.20
SIERRA LEONE	0.38
TOGO	0.40
OTHERS	0.12
AMERICAS	**1.39**
BRAZIL	0.21
COLOMBIA	0.28
DOMINICAN REPUBLIC	0.20
ECUADOR	0.28
VENEZUELA	0.25
ASIA AND OCEANIA	**2.11**
INDONESIA	1.60
MALAYSIA	0.31
PAPUA NEW GUINEA	0.10
OTHERS	0.10

REFERENCE: ICCO

Six Portraits of farmers

WORKING IN VERY distinct conditions and cultures these six growers all focus on producing quality beans. From the scientist focused on creating a bird sanctuary to the heir of a business adventurer in Madagascar, they have very different backgrounds and social statuses. But they are all convinced of the necessity of sustainability at all levels—agronomic, social and ecological—and they work toward it.

BERTIL AKESSON, AKESSON'S ORGANIC – BEJOFO, MADAGASCAR

CONTEXT

CACAO WAS INTRODUCED on the island of Madagascar as early as the nineteenth century using *Criollo* beans from Mexico that had transited via the Philippines, Java, and Ceylon (now Sri Lanka).

However, cacao farming only took off in the twentieth century under the impetus of two French pioneers, Mr. Millot and Mr. de la Motte Saint Pierre. Using beans coming from Java, the Millot plantation, founded in 1922, and the *Compagnie de Culture Cacaoyère (CCC)*, founded in 1924, developed large areas devoted to cacao farming. The two estates, that had a monopoly for cacao until the independence, still operate today respectively forty-two hundred acres and fifty-five hundred acres.

Shaken by the turmoil of Independence in 1960, and successive nationalizations and privatizations, these two plantations have become:

Plantation Millot - Following bankruptcy in 2004 due to speculating on vanilla, the business was taken over by two partners - Philippe Fontayne and Ykbal Yridgee. In 2016, the French specialty chocolate maker *Valrhona* took a forty percent share in the company.

Akesson's Organic Madagascar, owned by Mr. Bertil Akesson, Jr., is the largest producer in Madagascar (three hundred tons per year).

Beside farming cacao on its land (about two hundred tons per year), the Millot plantation collects a large volume of beans (four to six hundred tons per year) from small farmers working in the lower Sambirano Valley.

HOW DID YOU START IN CACAO?

BERTIL A - In 1976, *Compagnie de Culture Cacaoyère*, initially created in 1924, merged with another Malagasy agricultural company called *CAIM*, active in sisal and tobacco, and formed *SOMIA*. The 1977 military coup installed in Madagascar a socialist government that nationalized many companies, including *SOMIA*.

In 1979 my father, a Swedish businessman, took over *PIM* – a holding company regrouping various mining and agricultural activities in Madagascar, and previous owner of *SOMIA* before the nationalization. Twenty years later, the Malagasy government requested that he buys back *SOMIA* or lose some sisal titles belonging to the holding and linked to *SOMIA*. This is how, somehow by luck, he added cacao farming to his activities.

In 2004, I had already been working four years in the family company and I start marketing *SOMIA's* cacao. I discovered the potential of Madagascar cacao, which nobody really wanted back then, except for two French chocolate makers—*Valrhona* and *Pralus*—who made it their signature taste and were getting their cacao from *Millot*. This large potential was a revelation to me.

But in 2006, *SOMIA* was transferred to a third party, and *SAGI* was formed to manage exclusively the cacao activities belonging to SOMIA. This is when I left the family company, and decided to focus exclusively on cacao and – via my own company - *Akesson's Organic* - I started to negotiate the purchase of *SOMIA's* land where cacao was growing. Ultimately, I took over that land and rebranded it *Bejofo Estate* (Bejofo is the name of our farm/village where all our activities are centralized).

THE BOJOGO ESTATE

HOW DID YOU EXPAND?

BERTIL A - Having worked with my father over seven years on the ground, I knew Madagascar and its people well, although my activities were mostly related to sisal plantations and mining graphite and mica.

Back then, Malagasy cacao had the reputation to be rather acidic and generally poorly processed. The market wanted a cacao with strong chocolatey flavors and the least acidity possible. So I decided to learn about the post-harvest treatment – visiting cacao farmers in Venezuela and elsewhere – and improve it, enhancing the fruity acidity. I capitalized on Madagascar's difference and on our history. Then, of course, the timing was good. I was there when the American craft chocolate makers emerged and most of them actually started to make their first chocolate with my beans. The American market was absolutely key in our development. Then the craft trend "contaminated" the rest of the world as well and the rest is history.

THE HEIRLOOM PODS

WHAT IS YOUR SITUATION TODAY?

BERTIL A - Today, I sell beans to approximately two hundred chocolate makers. That is nearly all the high end *chocolatiers* of the world, from the French *Pralus* and *Alain Ducasse*, to the American *Dandelion* and *Amano*, the Chinese *Fuwon*, and the Singaporean *Fossa*. In Madagascar, to produce about three hundred tons of beans, I have a permanent staff of one hundred and fifty plus seasonal workers at harvest times, which could number up to seven hundred people. At the level of quality and price that we sell, I think we have all the market segment that we could possibly have. So our challenge is to grow this high quality specialty market segment. I would describe it as the craft and artisan *chocolatiers* who can both take advantage of the aromatic qualities of my cacao and who can afford it. Seeing the explosion of "bean to bar" makers, the trend seems positive. However, the relatively large number of newcomers does not translate into significant new volumes. This is why I am diversifying my crops by adding spices (black & pink pepper), vanilla, and other agricultural products specific to tropical agriculture and to Madagascar.

WHAT IS YOUR APPROACH TO CERTIFICATIONS?

BERTIL A - In Madagascar, I have been certified organic since 2005, I have the sustainable certification from *For Life* since 2016 and in 2017, my beans received the *Heirloom* recognition which guarantees their ancestral origin. In my case, that meant my variety can be traced to the authentic Mexican *Criollo* seeds. My plantation in Brazil obtained the sustainable certification from *Utz*.

However, I think certifications in general have a rather limited impact on the lives of the people on the ground. I came to that conclusion by observing that even in some organic-certified areas, there may be pesticides or chemical compost. Similarly, I have not identified any significant wealth difference between standard farmers and those who have been certified *Fairtrade* or *Utz*. This may be due to the fact that most of the cacao growing countries have legal institutions and overall cultures that have a different understanding of the rule of law. Sometimes rules and regulations are perceived as long-term targets rather than compulsory commitments. This is often compounded by enormous wealth gaps between the beneficiaries of the certifications and the people granting them. These behaviors are so engrained into some local cultures that they are not necessarily seen as illegitimate.

There is no reason to believe that the rules for being organic get a different treatment than the other laws and rules applied in the society at large. In effect, despite twenty years of certifications or maybe more, the situation for farmers on the ground has not changed much. I conclude that the main *raison d'être* of certifications is to respond to a real marketing need experienced by chocolate makers and retailers.

But even if the results are faint, there are a few, and it is a positive thing that the topics of organic farming and sustainability be permanently at the forefront in all aspects of the cacao industry.

THE LARGE FERMENTATION BOXES

MONEY

BERTIL A - All my activities are self-sustained and profitable. Obtaining quality cacao beans requires a lot of labor; so labor costs are a key component for the sustainability of a plantation. This is the main reason why when the cost of labor in a cacao growing regions goes up, which is obviously a good thing for the local standard of living, farmers tend to move away from cacao to grow alternative crops perceived as more immediately rewarding. These usually include fruit, rubber trees, coffee, cane sugar, etc.... depending on the soil characteristics.

Madagascar does not have the important influx of tourism experienced by many cacao-producing countries in the Caribbean, the Philippines, Vietnam, and other areas. This reduces the benefits of ecological tourism–which is sometimes an essential complement for farmers.

SORTING BEANS AFTER DRYING

OTHER TOPICS

BERTIL A - I am of course delighted by the current expansion experienced by the specialty chocolate and aromatic chocolate segments. However, as eighty percent of the world's cacao is made of commodity beans, this growth only affects twenty percent of production—meaning it ignores four-fifths of the market!

This is the market experienced by the people who buy chocolate, essentially in the Northern hemispheres, and by the millions of farmers growing cacao. Among this twenty percent of the world harvest, a minute share is of high quality, like my beans—certainly, less than ten thousand tons. Out of a world production in excess of 4.6 million, this is extremely slim.

Because of their size, these percentages are not going to change quickly. By comparison, it took more than a whole generation for American micro-brewers to reach a fourteen percent share in their market, whereas beer making is a much simpler process than chocolate-making and does not rely on imported raw material and expensive equipment. Coffee may seem a more optimistic example. This is a truly mass-consumed product, much more than chocolate, which has undergone major consumption changes over the last generation. However, it took eighteen years for the specialty coffee segment to grow from fourteen to forty-one percent! I do not forecast such a spectacular progression with cacao.

JEAN-YVES BRANCHARD, ANANDA COCOA – YANGON MYANMAR

CONTEXT

MYANMAR, THE COUNTRY formerly known as Burma, with sixty million inhabitants, is not currently known as a cacao country.

The national economic plan devised by the socialist, military government ruling the country in the 1970s included the introduction and development of cacao farming in various regions. Cacao seeds were imported and some farmers started cultivating the *Theobroma Cacao* variety in areas east of Yangon, towards Thailand. To further signal the country's commitment to cacao farming, Myanmar became a member of the *ICCO*.

However, without the necessary skills for post-harvest treatment (fermentation) and no commercial follow-up, results did not appear swiftly and the project was quickly deemed unsuccessful. The ensuing governments gave-up on the culture of cacao–which disappeared almost completely.

In recent years, witnessing the success of neighboring countries like Vietnam and, to a lesser extent, Thailand, some Myanmar government representatives have voiced their support for cacao farming. But effective on-the-ground measures are hard to find and the Burmese cacao crop is currently minimal, estimated to be below one hundred tons.

DETERMINATION WINS

HOW DID YOU START THE CACAO ACTIVITY?

JEAN-YVES B - In 2003, as I was looking for new activities, I decided to try growing cacao trees. Aware of the problems Vietnam went through when it started cacao farming I was very concerned about diseases, and was adamant to obtain disease-free seeds and robust varieties. I contacted the French Agricultural Research organization for Development, *CIRAD* in Montpellier. After various back and forth conversations and following their advice, I received three thousand cacao seeds via DHL from a farm they had in Vanuatu!

I selected two different areas. The first one in the Karen region, northeast of Yangon by the Thai border. This area was sporadically affected by guerrilla activities, but the existence there of some of the old cacao trees from an abandoned 1970's project indicated that the

CACAO SEEDLING PROTECTED BY THE "BETEL" TREES

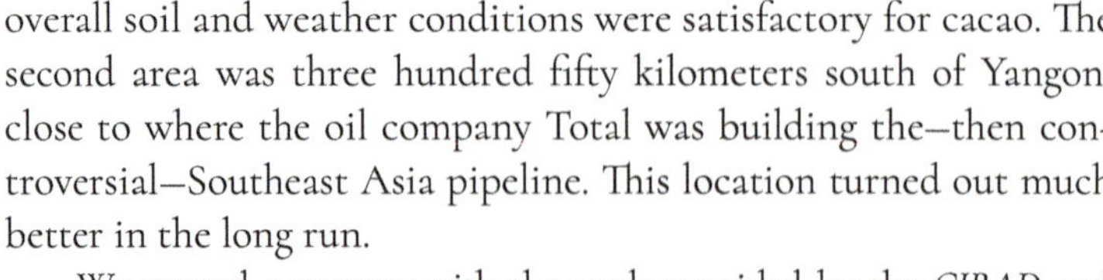

overall soil and weather conditions were satisfactory for cacao. The second area was three hundred fifty kilometers south of Yangon, close to where the oil company Total was building the—then controversial—Southeast Asia pipeline. This location turned out much better in the long run.

We started a nursery with the seeds provided by the *CIRAD* and eventually planted seedlings. However, the area set aside for the cacao plantations needed some adaptation because, among other characteristics, this region does not have enough nebulosity to prevent evaporation of humidity. Extra shading is therefore required to maintain a rainforest atmosphere. Fortunately, a local tree extremely common here for its fruit, the *Betel* (*Areca catechu tree*) has the ideal characteristics to provide both shade and soil nutrients. Another negative factor is the heavy rains during the wet season pouring between two to six meters of water overall.

After our efforts, the internal war came back and we could not access the land during three years. Then about twenty-five acres (about ten hectares) burned following a lightning strike on nearby rubber trees.

HOW DID YOU EXPAND?

JEAN-YVES B - In 2012, we relaunched the cacao project in partnership with farmers in areas no longer under a state of war. In Myanmar, there is a lot of competition for accessing land. Despite the highly fluctuating price of latex, rubber tree farming remains attractive and, of course, the skyrocketing demand for palm oil also generates pressure on land availability. The cultural habits here also create another burden for anyone wanting to acquire land in Myanmar. Unless your transaction is implemented with total transpar-

TRAINING FARMERS

ency and with public blessing from the village chief, neighbors and sometimes the local shaman and the sorcerer, your deed could be challenged and you could not be the owner of the land after all.

We replicated our first approach by buying new seeds from *CIRAD*, creating nurseries, and involving farmers from day one.

As there is no cacao culture here, I started training farmers on cacao grafting, pruning and harvesting. I also built a facility to ferment the beans on my own. Back in Yangon, I created a small chocolate laboratory to produce quality chocolate bars at different percentages. I was new to chocolate making and learned by reading as much material as I could find online and in books. I also experimented many, many times in a trial and error process. For example, we bought a small Pomati tempering machine and it took us some time and effort to fine-tune it to the viscosity of our two-ingredient chocolate.

For a couple of years now, our chocolate production has been selling in many international hotels in Yangon, at airports and at some other tourist locations in the country. However, I am aware that some improvements in quality and productivity are needed in the chocolate lab.

WHAT IS YOUR PRESENT STATUS?

JEAN-YVES B - The nascent notoriety I have gained by being the first chocolate maker in Myanmar helps me to increase sales in a growing number of outlets. I even received proposals from the Myanmar government to represent the country at international cacao events! I see a lot of interest for both cacao farming and chocolate making. Farmers like the prospect of harvesting a crop that seems to require

YOUNG CACAO PODS IN THE SHADE

little to no maintenance. But without knowledge and proper care, there will inevitably be a negative impact on quality and this will taint the reputation of Myanmar as a cacao origin. I believe the worldwide demand for quality cacao beans will continue to grow and, as a small producer, Myanmar should target the high quality segment because it offers better margins and would allow farmers to avoid a losing competition on price with large growers from West African countries, and from Indonesia or the Philippines.

I currently operate two plantations with a combined size of about sixty acres (about twenty-four hectares). I work exclusively in agroforestry with a low density of cacao trees (six meters minimum between trees). Note that the main shading trees - the areca tree - are harvested once a year and do not need chemical or pesticide maintenance. Cacao plantations intercropped with areca allow farmers to easily double their income.

My goal is to have approximately thirty thousand cacao trees within two years and to operate a full size chocolate factory in five years. I do realize that meeting this goal requires a lot of changes in my activities as well as in the rural management of the country, for which a stable political environment is a pre-requisite.

WHAT IS YOUR APPROACH TO CERTIFICATION?

JEAN-YVES B - I do not have any certification. I am not even sure if the various certification organizations operate in Myanmar. Once I am able to export cacao and/or chocolate I will probably be interested in the USDA Organic stamp.

From what I have seen in other countries, the certification process is not immune to political and financial interference and therefore I will only go through the expensive process if customers require it. For my own chocolate, I know how I work at the farm and therefore I know it is fully organic and sustainable.

For quality purposes, one certification or competition I would pursue is the one called *Cacao of Excellence*. It focuses on the actual fruit and its qualities as derived by proper farming, not on the chocolate process. I hope to be ready for it within two years.

MONEY

JEAN-YVES B - Until now, I consider the farming and chocolate-making activities as one business. I do not have enough beans for myself so I do not sell to anyone, internally or abroad. Taken as a whole, my cacao and chocolate business is self-sustaining–so much so that I am actively looking for ways to expand my crop to meet the chocolate demand. At the same time, I am also considering buying cacao beans from other countries to develop a full chocolate portfolio of flavors and origins.

OTHER TOPICS

JEAN-YVES B - Considering cacao in Myanmar, I have two main concerns. The first is the sustained, real and effective involvement of governmental instances. Because of its deep impact on rural lives and revenues, the decision to pursue cacao farming has political consequences that need to be taken care of. So far we have seen good progress in roads and water management, but much more could be (and must be) done to promote cacao and train farmers, to organize cooperatives etc.... as well as to bring awareness to the world that Myanmar is now producing quality cacao.

Secondly, is the quality of beans. Cacao diseases can spread quickly if farmers are not educated to cure the issue from its first inception. Without a widely spread cacao knowledge, an agronomical accident could rapidly degenerate. We must avoid repeating what happened in Vietnam and Malaysia in their early cacao ventures.

Having said that, I hold the Vietnamese chocolate company *Marou* as an example of where I'd like to be in the long run. This means to operate two well managed mid-size plantations of about twenty-five acres (ten hectares) where I would also collect a lot of other beans from neighboring farmers while making quality chocolate to be sold in my retail outlet as well as internationally.

DELIVERING SEEDLINGS DURING THE WET SEASON

CHRISTOPHER FADRIGA, PLANTATION DE SIKWATE – THE PHILIPPINES

CONTEXT

THE ARCHIPELAGO OF the Philippines has been growing cacao for more than four centuries. The Spanish colonists first brought cacao seeds from Mexico at the beginning of the seventeenth century. The original Mexican *Criollo* variety was grown in most islands of the colony. At the turn of the twentieth century, when cacao demand started to explode, and the European buyers were establishing large productive plantations in West Africa, the large land owners of the southern island of Davao joined in. They planted similar high yield and resistant varieties to the ones used in Ivory Coast, Ghana or Cameroon, all belonging to the *Forastero* category.

Today, the Davao province produces over eighty percent of the Philippines' cacao. Nearly all of it is sold as *Commodity*. Recently, responding to the growing demand for aromatic beans, farmers in Davao and from other islands, are exploring fine aroma varieties.

Christopher Fadriga is one of them.

FROM FLOWERS TO CACAO

HOW DID YOU START THE CACAO ACTIVITY?

CHRISTOPHER F. - In 2013 I had been growing flowers forty-five kilometers from Cebu City for many years and was already committed to organic agriculture. One day in the fields I found a cacao pod that was very soft-skinned and that intrigued me. It was so soft that I could crush it open with my fingers. That impressed me a lot because I had never touched such a soft cacao pod, plus, it really smelled good. I posted photos on Facebook and Steven DeVries, a chocolate fanatic and bean to bar maker in Denver, Colorado (USA) contacted me for details. The Japanese food importer and wholesaler *Tachibana* also asked questions. Both suggested it might be a remnant of the old aromatic Criollo, believed to have disappeared, and initially brought to the archipelago from Mexico by the Spanish colonists in the seventeenth century. *Tachibana* also mentioned that if it were the case, there would be a high demand for it in Japan as consumers were increasingly interested and buying aromatic chocolates.

In 2014, a good friend of mine, Mr. Mel Santos, invited me to a four day seminar in Malagos Garden, sponsored by the American food giant Mars to teach farmers how to properly grow cacao. During an open session I talked about the likely presence of the old aromatic Criollo on the archipelago and about the potential for Fine Filipino cacao. Most of the participants were growing bulk beans sold to industrial food groups like the organizing company, and were not interested in any future crop without a clearly defined paying customer. I did not attract a lot of interest. However, this was the crucial moment for me because a handful of farmers were enthusiastic about the idea.

CHRISTOFER FADRIGA
AUTHOR ALAIN d'ABOVILLE
MEL SANTOS

Mr. Santos, who was growing hybrid varieties of cacao on land owned by the religious order Brothers of the Sacred Heart in Davao, and I became very interested at the possibility of having an aromatic variety in the Philippines. We launched a social media campaign asking Filipino people to look in their backyard, their gardens and in the open fields, to find cacao pods that had a thin skin and pale colored beans. It was an encouraging surprise to receive many photos of potential old *Filipino Criollo*. We went to check the most promising trees to gather seeds.

HOW DID YOU EXPAND?

CHRISTOPHER F. - In 2015, I started a nursery in Atipulan Bago City, on the Island of Negros to produce mother trees of the white bean cacao trees using the seeds we had collected through the online campaign. We also spread mother trees through the cacao growers' *Association Plantation de Sikwate*. The goal was to be able to produce enough grafting material to allow new farmers to grow the pale bean variety which by then we started to call *Filipino Aromatico*. After three years, we now have enough mother trees to cover our needs for grafting material—which we are sending all over the archipelago.

Now, the Japanese food importer *Tachibana* is helping us to ascertain the genetics of the *Filipino Aromatico* bean. As cacao is a famously promiscuous plant, it is unlikely that after so many years we obtain a pure *Mexican Criollo* as there is probably no such thing anymore on the planet. However, we have identified many aromatic trees with pale-to-white beans that have triggered a lot of interest from Japanese and European chocolate makers.

WHAT IS YOUR PRESENT STATUS?

CHRISTOPHER F. - The *Plantacion de Sikwate Cacao Producers Association* has been very successful by heavily relying on social media. Today the largest *Filipino Aromatico* project is being implemented on the Island of Negros near Bacolod Airport. On a vast fifty-two hectares farm, previously dedicated to sugar cane farming, we are planting forty hectares of *Filipino Aromatico*. After uprooting all the cane, we planted twenty-five thousand banana trees to provide shade for the young cacao trees as well as some other vegetation to control the burst of wind sometimes happening in this area. In order to be organic, we have created a compost station composed of large worm pools filled with manure and cane mud press. We poured nine hundred tons of sugar cane mud cakes per hectare. The effect is so strong that after only seven months, the banana trees are already bearing fruit. To prevent the damage of seemingly longer dry periods, we are building a dripping irrigation system getting water from deep wells using solar powered pumps. This year we are entering into the planting phase. The attached nursery has now enough material to progressively plant the forty hectares. Cacao harvest is expected after two to three years.

WHAT IS YOUR APPROACH TO CERTIFICATION?

CHRISTOPHER F. - We are organic, this is our DNA. For the same reason we are taking good care of the people involved in our activity, not only because we are good people, but also because only happy workers produce good products and can take care of their community. So, I believe we qualify for any certifications, be it organic, Fairtrade and others. Because we are catering to a niche market that gets involved in the supply, we do not feel the need to support the financial cost of certifications. For example, our potential Japanese customer has visited our plantations many times and we provide him with the information he requires regarding our use of organic compost and organic pesticide. He is involved in defining the post-harvest protocol, etc. So, in a way he is "self-certifying", but we are not getting into it ourselves.

VISITING AT THE OPENING OF INDAI LE CORTES'S PLANTATION

MONEY

CHRISTOPHER F. - Farmers, in general, are not getting enough for their cacao beans and I will always find the price paid to local cacao growers too low. Here, the current price for fermented dried beans is around hundred pesos per kilogram (*Editor's Note: about US$2, the commodity price*). In Davao and elsewhere there is a trend asking the farmers to perform the fermentation themselves instead of going to the large central post-processing centers. To be well done, the fermentation process requires some technical skills and volumes that individual farmers usually lack. But everyone is eager to climb the value chain and, of course, they get more money than from selling the wet beans.

Another trend to increase revenue is to produce chocolate products. I think this can only be effective if there is a local market as small producers do not have the resources to sell nationwide and indeed to export. In the Philippines, there is a huge market for the locally produced *Tablea* chocolate. It is a chocolate bar made by grinding beans, not necessarily fermented, and molding them in balls that people melt in hot milk or water to produce drinking chocolate. This is a product that could be developed by growers, probably by uniting their efforts inside a cooperative or some other associative structure.

CHRISTOPHER FADRIGA'S NURSERY ON CEBU ISLAND

OTHER TOPICS

CHRISTOPHER F. - From my experience in cacao farming, I believe there is a growing niche market in search of farmers capable of producing quality, aromatic cacao beans. Moreover, when there is credible, meaning true, story around it, it adds value. The *Filipino Aromatico* can be traced back to the Spanish Galeon Santisima Trinidad, the ship that brought the Mexican seeds in the early seventeenth century. This creates a link with history and with people. I am sure this will help us reach our goal to have ten million *Filipino Aromatico* trees in ten years and spread their good aromas all over the world.

MONICA LILIANA, MARIANA COCOA EXPORT – SANTANDER, COLOMBIA

CONTEXT

FOLLOWING DECADES OF international branding, Colombia is world renowned for its coffee, a plant that was imported from East Africa a couple of centuries ago. Cacao is indigenous to the western regions of Colombia where original varieties can be found and the plant is grown in most provinces. Cacao has been cultivated for centuries but, until recently, was underperforming because of the internal turmoil created by the unstable political situation. Since the peace agreement aiming to re-integrate internal violent dissidents into society, Colombia has identified cacao as an essential tool to spread peace and prosperity to rural areas. Many international development organizations - USAID, the PNUD, the European Union, and some European countries like Germany, France, etc....- participate in programs such as *Cacao for Peace* whose objective it is to train farmers to grow cacao instead of coca.

Colombia's international reputation for chocolate is constrained by its internal consumption. The historic national giant *Casa Luker*, created in 1906, controls most of Colombia's harvest. Until 2012, there was no surplus over the forty thousand tons needed to satisfy the local demand for chocolate. Now producing more than fifty-five thousand tons and growing (2017 figures), Colombia has spare capacity and is increasingly active on export markets.

Local consumption is mainly satisfied with bulk beans, and exports are nearly exclusively made of aromatic cultivars (ninety-five percent). The overwhelming majority of the harvest is composed of productive and commodity beans, including *CCN-51*, acquired by large corporations like *Casa Luker*" and by a few international buyers. However, the country holds many aromatic heirloom cultivars and indigenous varieties and hybrids of high quality.

In order to improve growers' revenue and community life, some institution are promoting aromatic varieties. *Mariana Cocoa Export* was one of these pioneering structures at its launch in 2003 and remains a leading force in Colombia's quest for quality cacao.

ONCE UPON A TIME, THERE WERE FOUR SISTERS.

HOW DID YOU START THE CACAO ACTIVITY?

MONICA L. - We, the four sisters, Mariana, Marcella, Monica and Paola formed *Mariana Cocoa Export* in 2003 using the family business started by our agronomist father as a springboard. Exploiting historic connections with Ecuadorian and Venezuelan customers, *Mariana Cocoa Export* was able to quickly export conventional cacao beans from these countries. When the company approached the European market, I realized there was no demand for commodity beans from Colombia, as EU buyers are already satisfied with West African and Indonesian products at lower prices. It also became clear to me that there was a growing interest for aromatic beans—provided they had undergone a correctly executed post-harvest treatment.

To be able to produce and sell such crop, one of us, Monica in effect, was dispatched to study the necessary topics in Ecuador so that she could ascertain the quality of the production and spread the knowledge via training programs to Colombian cacao farmers.

THE PRIZED AROMATIC BEANS

HOW DID YOU EXPAND?

MONICA L. - Between 2006 and 2007, *Mariana Cococa Export* started its first training session in some of Colombia's cooperatives. For historical and logistical reasons we do not have large post-harvest centers performing the fermentation, drying, storage and packing of cacao in Colombia. I can only think of one in the Tomato province. In Colombia each farmer operates his/her own small fermentation boxes and dries beans wherever possible. This creates an issue for us to guarantee the quality of the end product, and a need for on-the-ground tailor-made training. Through extensive training and constant oversight (we test/control over three hundred cacao samples each year), we've been able to maintain the highest possible level of post-harvest treatment. This allowed us to start selling in Europe. More recently, we started participating in international chocolate fairs, such as the *Salon du Chocolat* in Paris and the Harvard Square Chocolate Festival, to make us visible to buyers. These operations are very useful, not only to put us and Colombian cacao on the world stage and meet potential cacao buyers, but also to evaluate the market, to identify new trends, and assess the competition.

WHAT IS YOUR PRESENT STATUS?

MONICA L. - We currently work with multiple associations, maybe around forty, in four provinces—Arauca, Putomayo, Santander, and Antioquia. Each association groups between fifty and one hundred fifty farmers, mostly of small size. It is important to mention that the average age of cacao farmers here is high, most of them are in their late fifties to mid-sixties. So, we do everything we can to motivate their kids to work with their parents and continue the cacao farm. The main motivational triggers are lifestyle - decent housing, school and community life - and of course revenue.

Our training program is based on four components:

- *Sensorial / Quality* - The organoleptic properties of cacao along its advance in the post-harvest process, and indeed in the chocolate it eventually produces.
- *Technical / Productivity* - Plantation characteristics, soil quality including pH levels, irrigation, wind. etc. The available cultivars, running a nursery, grafting, etc....
- *Gastronomy* - With the objective of creating touristic projects, a module on traditional cooking and chocolate preparation is taught.
- *Commercial* - How to present cacao beans, organoleptic descriptions, competitive tasting. How to define customers, identify and approach them.

We constantly run training courses and adapt them to the specificity of the local audience.

In terms of cacao exports, last year was not our best as we only exported three tons of aromatic fermented dried beans. We are expecting a swift increase in volume as the training programs are ramping-up, which will vastly improve available volumes as well as the quality of the crop.

WHAT IS YOUR APPROACH TO CERTIFICATION?

MONICA L. - I think it is now clear that we are looking first and foremost for quality. With this in mind, we also seek recognition of our work through certification. Some of the cooperatives we work with are *Utz* and organic certified. However, many small producers do not have the volume and therefore the financial resources to undergo the certification process. Sometimes external partners such as USAID and others co-finance these operations. This is of growing importance to us because some markets like Canada, for example, will simply not buy beans without a recognized economical (like *Fairtrade*) and/or agronomic label (organic). So yes, we recognize the value of certification as some of them are a pre-requisite to access markets. This is a major challenge for us, *quality seekers* because only a small fraction of Colombia's cacao has at least one certification of some kind. I just wish buyers requiring them would participate more directly in their financing or accept a higher premium price for it.

MONEY

MONICA L. - All cacao professionals are aware that cacao prices paid to the small farmers are too low. This situation is compounded in Colombia by the volatility of the Colombian pesos against the US dollar. Commodity prices are set in US dollars and the evolution of the currency adds more ups and downs to the already low and fluctuating commodity prices used as reference when fixing the compensation given to farmers. Small farmers are having a hard time getting by selling their commodity beans for two thousand pesos per kilograms (*Editor's Note: about US$0.60*). Beans from growers applying our strict quality criteria are bought for more double that price at five thousand pesos (*about US$1.50*).

As *Mariana Cocoa Export* we've been afloat for many years thanks to our consulting and training activities.

PROMOTING THE MANY GREAT CACAO ORIGINS FROM COLOMBIA

OTHER TOPICS

MONICA L. - Our paramount goal is to establish and promote Colombia as a producer of quality cacao. Already most of the country's exports are *Fino de Aroma* which is a good thing. But we, as *Mariana Cocoa Export*, want to increase both the quality and the volume of Colombia's cacao exports. When attending international fairs, we are sometimes surprised that fine *chocolatiers* do not know Colombia is a producer of aromatic cacao. Colombia does not come to their mind when searching for new quality origins. We want to fill this deficit of information and make sure that every chocolate professional in the world is aware of the quality of our trees. There is no reason we cannot be like, or better than Ecuador or Venezuela now that our political situation is stabilizing.

DR. SANH, MARKRIN – CHIANG MAI, THAILAND

CONTEXT

THAILAND IS NOT a member of the *International Cacao Organization - ICCO -* and is not known as a "cacao country" although cacao seeds were first introduced from Malaysia and Java in the south of the country around the 1920s as a commodity product. But rubber plantations were more profitable and cacao farming did not expand vastly. Because of the current market conditions and of a local political will, cacao is now experiencing a significant revival. The country produces less than three thousand tons harvested by about two thousand farmers, mostly in the southern regions, but also around the northern city of Chiang Mai.

The resurgence of cacao farming started less than ten years ago (in the early 2010's) and there is a growing interest from many members of the communities. As revenue from farming natural rubber is decreasing some hevea plantations are starting to grow cacao alongside their rubber trees. The influx of western tourists looking for quality and authentic products is also a factor in creating a local demand for chocolate and touristic activities. Entrepreneurial farmers are seizing the opportunity to launch cacao on their land, together with eco-tourism projects. Ultimately, the trend of chocolate consumption in Asia shows that close neighbors like China and Japan could soon become important buyers of Thai cacao and chocolate, turning the tropical fruit into a very lucrative activity.

Among the many varieties grown in Thailand, the most common is the *Chuprun* which is particularly well-adapted to the meteorological conditions of the south of the country but does not behave as well in the northern regions where climate is both wetter and drier at times.

Because of its capacity to sustain communities in rural areas, the government and many social organizations are supporting efforts to create an indigenous cacao industry.

With his *MarkRin Chocolate and Farming* businesses, Dr. Sanh is both a leader and a major actor in this current evolution.

HOW DID YOU START THE CACAO ACTIVITY?

DR. SANH - Thirty years ago as a professor at the horticultural Maejo University in Chiang Mai, I started working on coffee, tea, and cacao. I quickly became convinced of the potential of cacao in Thailand, and I started a research program with my wife to investigate how to develop this plant and help its spread in Thailand.

About twenty years ago, during an exchange program with the American University of Wichita, in Kansas, we received a professor who brought some cacao seeds from Peru. I crossed this variety with a Filipino bean and produced a new hybrid which I called *I.M.1*, using the initials of my children, Iran and Mark. I registered and patented the *I.M.1* variety which is only produced by *MarkRin*. It is made of *Criollo* and *ICS* from Peru, and of *Forastero* from the Philippines, and we are the only ones cultivating it.

Around 2006, because of this achievement at creating a hybrid, I decided to grow cacao myself in order to have a full scale experience for my students. At the time, we were the first and only cacao grower in Chiang Mai. We now have agreements with farmers to whom we provide the seeds and we buy their harvest in order to make our chocolate and by-products (such as cacao liquor and butter).

The high yield of the I.M.1 is due in part to the fact that its flower carries the male and female organs in closer proximity than other varieties. This vastly improves self-pollination compared to other varieties which, in turn, results in an extremely high percentage of flowers becoming cacao pods.

PROMOTING THE IM1 VARIETY

HOW DID YOU EXPAND?

DR. SANH. - In 2008 my wife and I started *MarkRin Chocolate* to test and sell the chocolate made with *I.M.1*. We initially started with a simple mortar and pestle, and then we bought some specialized equipment.

We initiated a promotion plan for our bean and it is currently grown as a single crop, mostly but not at all exclusively, in the Chiang Mai area and also as part of intercrop farming. In Thailand, cacao is most often cultivated alongside rambutan, bamboo, rubber or lychee. In these instances, the use of chemicals for these other crops prevents the cacao from being organic. So only the mono-crop plantations qualify as organic.

We are exclusively self-financed, so we advance step-by-step at a pace we can afford. We have many projects for increasing both the spread of the *I.M.1* variety and the chocolate production.

MRS. SANH IN THE LABORATORY

WHAT IS YOUR PRESENT STATUS?

DR. SANH. - We now have about five hundred thousand *I.M.1* trees all over the country. It is interesting to note that although farmers are using the same variety country-wid, our *I.M.1*, the aromas of the chocolate made with beans from different regions are different, resulting in distinctive tastes and flavors. The chocolate is definitely nuttier when using cacao beans grown in the north and fruitier when coming from the south. Such is the influence of the terroir.

We now produce a lot of chocolate and we are in the process of planning new factories, first in Chiang Mai and possibly another one in the south of Thailand in a more distant future.

WHAT IS YOUR APPROACH TO CERTIFICATION?

DR. SANH. - We understand the need for certification and want to obtain a few key ones. We started with the organic classification from the *US Department of Agriculture - USDA Organic* - We are at the end of the comprehensive and lengthy certification process and *MarkRin Farms* and *MarkRin Chocolate* are likely to be USDA Organic Certified within months. The next step will then be to work with farmers, to help them maintain the classification. After that, we will work on obtaining an Utz Certification.

Unlike West Africa and some parts of South America, there is no child labor in Thailand. However, not everything is perfect, as low income for farmers and rural activities in general create unsustainable conditions. We hope certifications such as *UTZ* will help.

MONEY

DR. SANH. - Concerning our own businesses, *MarkRin Farms* and *MarkRin Chocolate*, we are indeed self-sustaining, and we are self-financing our development. Concerning the farmers, we work with, we consider that their revenues are not satisfactory because of the pressure on price to the expense of quality and sustainability. As our *I.M.1* is highly productive, up to five tons per hectares, and with low acidity and high fat content, farmers do obtain a better price than with the *Chuprun*. In general, cacao yields are very high in Thailand because of good pruning and overall excellent maintenance of the plantations. As *MarkRin Chocolate*, we produce tailor-designed chocolates for our farmer's customers providing them with extra revenue. The customization is centered on the fermentation process which is how we deliver differentiated aromas.

IM1 CACAO PODS

Finally, we encourage farmers to develop eco-tourism activities whereby visitors pay to learn how cacao is grown and chocolate made. During their visit, tourists do consume and buy products from the farm including chocolate.

OTHER TOPICS

DR. SANH. - It has only been five to seven years since Thailand dedicated resources to cacao farming and the current growers are all new to the crop. Their number is growing steadily because they obtain a good revenue and, via intercrop plantations, they can start while continuing to grow their traditional plants. This allows them to continue their existing farming while learning about cacao.

I believe that eventually, and not in the distant future, Thailand will become a member of the *ICCO*. Thailand meets all the necessary criteria to produce large quantities of quality or aromatic cacao: a suitable weather and soil, an educated workforce, a good transportation infrastructure, a growing indigenous market, and large export markets nearby. So, I predict a bright future for Thailand cacao!

CHARLES KERCHNER, RESERVA ZORZAL – DOMINICAN REPUBLIC

CONTEXT

DISCOVERED BY CHRISTOPHER Columbus in 1492, the island of Hispaniola, which is now shared between the Dominican Republic (D.R.) and Haiti, started growing cacao at the end of the sixteen century using beans from Mexico. This was many years after the Spanish had imported sugar cane, now the first agricultural production of the island.

A historical figure of Caribbean cacao and chocolate, Pedro Cortez, started large scale production in 1929 and gradually moved to Puerto Rico where the company is now based. However, because of the high cost of labor in the American territory, Cortez now imports most of his beans from the D.R. where cacao farming now represents a production of nearly seventy thousand tons (2018) of which one-third is fermented and therefore labeled *Hispaniola* and the rest is denominated *Sanchez*.

Recently, the D.R. has decided to make cacao a key instrument to help rural communities and is devoting resources to support cacao farming with the goal of tripling productivity by 2050—without using the dreaded CNN-51 variety. This National Action Plan affects primarily the traditional producers exploiting large plantations and selling at commodity price to the big international companies. There is also support for the growing number of farmers focusing on organic and agroforestry culture using the traditional high-quality varieties introduced centuries ago from Venezuela and Central America. The Dominican Republic claims to be the largest producer of organic beans in the world. However, exact quantities are hard to find and range from five-to-ten thousand tons (2017). A small number of chocolate makers on the island turn six percent of the harvest into chocolate sold locally to tourists and Dominicans.

With its wide bio-diversity and relative ease of access, the Dominican Republic is becoming a magnet for the development of new cacao plantations that are committed to being organic and sustainable while growing quality/aromatic varieties.

As the first ever Private Reserve on Hispaniola, *Zorzal* is a unique example of this new breed of farmers.

FROM BIRDS TO PODS

HOW DID YOU START THE CACAO ACTIVITY?

CHARLES K. - I initially came to the city of Nagua, on the Atlantic coast, in 2001 as a Peace Corps volunteer working on agroforestry. After my time on the island, I embarked on a graduate school program followed by a PhD. I published my thesis in 2012 on the conservation of migratory birds in North America. I realized that these animals, in particular the small Bicknell Thrust, were losing their natural winter territories in the Caribbean due to growing urbanization and extended use of chemicals in farming.

To provide them a sanctuary, I decided to start a reserve where they could find the pristine environment they need. Because it is already a significant winter retreat for birds and because of my experience there, the Dominican Republic was an obvious choice for me.

The Dominican government had recently passed a law authorizing the creation of private reserves, so in 2014 with resources from the Moreno family and a New York Foundation we established the first private natural reserve in the country on a thousand and eighteen acres. We called it *Zorzal* because it is the name of the Bicknell Thrust in Spanish. In order to be able to maintain the bird protection program over time we needed some stable revenue, and this is when we decided to grow cacao, using the organic and agroforestry methods that I had studied.

ROAD SIGN DIRECTING TOWARDS ZORZAL'S RESERVE

HOW DID YOU EXPAND?

CHARLES K. - Some of the land had recently been used for cattle, therefore it was a gradual work to grow a homogenous territory. To fully adhere to the agroforestry system we planted a wide diversity of plants and fruits, such as pineapple, mango, pomegranate, and even coffee.

With time the farmers beyond our reserve became interested in what we were doing. We now monitor birds on their land and provide them with cacao seeds and advice. We also buy their cacao crop and brand it separately from ours as Zorzal Community Cacao.

The Dominican government has developed a plan to help cacao farming on the island. There are three thousand acres designated in this plan where we could potentially expand. But we are not growth driven but quality driven.

WHAT IS YOUR PRESENT STATUS?

CHARLES K. - We produce about one hundred tons of fermented dried cacao beans with a permanent staff of seven, which grows significantly at times of peak harvest. Our five-year goal is to reach two hundred tons. At that stage we will have reached a big enough size. But this is not going to be easy. Our land is fairly inaccessible; in some areas we collect cacao using donkeys to carry the pods to the fermentation center. Additionally, the effects of climate change are every year more noticeable. We clearly experience drier and longer dry seasons and the rainy seasons definitely bring heavier rains than before. Until now, we have not had any major diseases on cacao trees, just a few cases of Black Pods which we were able to swiftly contain.

We consider ourselves as scientists with a mission, not as chocolate luminaries or revolutionaries. We are not here to change the world but to make people happy using technologies to improve their lives and the environment. This is why we permanently spend a lot of effort on improving our fermentation process, the drying method and every aspect of cacao farming. We also spend a lot of time on monitoring the birds and providing them with the necessary food and shelter they need, especially during the winter months when they are more numerous.

MANAGEMENT PLAN FOR THE PRIVATE RESERVE EL ZORZAL

STRATEGIC OBJECTIVES

- To preserve the humid forest and the territorial integrity, as well as to promote activities compatible with the area's exclusive goal of ecological conservation and restoration.
- To achieve acceptable levels of production, from an economic and ecological stand-point, in the area designated for sustainable agriculture.
- To meet satisfactory levels of complementary revenue from ecological services or other mechanism generating alternative revenue.
- To instill an harmonious and cooperating climate between the Reserve and the local communities.

WHAT IS YOUR APPROACH TO CERTIFICATION?

CHARLES K. - I think they are good overall. For example, *Fairtrade*, by imposing a minimum buying price for farmers, is going in the right direction and helps. And of course it is a good thing to recognize organic farming, etc. Here, we only have the *USDA Organic* certification that some of our customers asked for because it is a requirement if they want to sell their chocolate at *Wholes Food*, the American leader in quality food retailing.

Sustainability, both social and economic, is an essential *raison d'être* for *Zorzal* and we appreciate the work done by *Utz* and the *Rain Forest Alliance*. But our clients know us, they come to see us frequently, we have an open-book policy that allows them to be certain of the way we work. So we are not seeking other certifications because our buyers do not ask for them.

MONEY

CHARLES K - Thanks to the cacao sales, the reserve is self-sustained and can pursue its bird mission. Our mission requires that we save - meaning leave aside - seventy percent of the land under our purview. So we only grow cacao on thirty percent of the available land. And with the proceeds of that work we are able to pay all our Reserve and cacao farming staff, plus the various installations we need here and in San Francisco de Macoris where we have a warehouse.

OTHER TOPICS

CHARLES K - I am glad to see the increased demand for high quality cacao beans like ours in the market. But it remains a small fraction of the global cacao production. Most of the varieties in the Caribbean are of good quality but most of the cacao is not sold as quality crop. This is because most of the chocolate makers are not interested in paying for aromas.

I think therefore that a lot of education and information is needed so that end consumers realize the difference between a commodity chocolate, a quality bar and a specialty one, and become ready to pay in order to enjoy that difference.

THE "NEW" CHOCOLATE MAKERS

IN THE PRODUCING countries as well as in the traditional manufacturing regions, small, specialized, quality oriented chocolatiers are cropping-up. Taking advantage of new, smaller and cheaper equipment, and sharing chocolate knowledge online, these businesses produce quality cacao and seek close relationship and cooperation with the provider of their beans. The connection between Monica, running Carlota Chocolate in Giron, in the Santander province of Colombia, (two hundred eighty miles northeast of Bogota), Ben Rasmussen, crafting chocolate bars from his small laboratory in Woodbridge, near Washington, D.C., or Mr. Ho, hand-crafting bars in the back of his vegan restaurant in Chiang Mai, Thailand, is an obsession for aromas, a deep feeling of social responsibility towards cacao growers, and an unstoppable creativity to try to achieve better results without damaging the earth and its inhabitants.

Together they are the frontline of change towards sustainable quality chocolate.

Six Portraits of Bean to bar makers

MAXIME SIMARD, QANTU – MONTRÉAL, CANADA

CONTEXT

IN THE CANADIAN province of Québec, specifically in Montréal, the bean to bar *chocolatier Qantu* is operating in a rather small market which does not have an indigenous chocolate making history. In the mid-1990s the international chocolate behemoth *Barry Callebaut* opened a huge factory forty miles (sixty kilometers) east of Montréal and its sweet candy-like products are the chocolate reference for most people in the province. This factory is so large and produces for so many markets beyond Québec that chocolate products are in fact the second food export of Québec just behind pork!

Chocolate lovers in Montréal, and in Québec, are more used to sweet products than to dark chocolate. This requires the handful of local bean to bar makers to spend a lot of time educating consumers about flavors, aromas, and the many nuances and sensations present in craft bars made with quality beans. In theprovince of Québec, the commercial outlook is evolving rapidly, thanks to the renowned dynamism of Québecers. Now a significant and growing proportion of local chocolate fans is appreciating the many quality chocolates being imported in Montréal from the US and Europe.

Recognized internationally, *Qantu* has all the characteristics and skills to attract them to its own locally crafted aromatic chocolates bars.

HOW DID YOU START IN CHOCOLATE?

MAXIME S. - I met Elfi, my wife and business partner, about twelve years ago in Peru, during a trip to Machu Picchu. Our idea of starting a business together goes way back to when she was studying engineering and I was learning IT. Because of Elfi's Peruvian roots and the importance of coffee in her country, we initially looked at coffee opportunities in Québec. We realized that there were already over fifteen roasters (*torréfacteurs*) in Montréal and the market seemed already saturated. Simultaneously, we noticed the arrival of the bean to bar phenomenon in early 2010s and the evolution in the chocolate market. This opened our mind and palate to chocolate. After I had completed an MBA program in 2016, we did a brief tour of Asia and visited the famed chocolate maker *Marou* in Vietnam. This sealed our idea of getting into cacao, and we decided to study and, more importantly, to try making our own chocolate bars. Our first experiment with two kilograms (approximately 4.5 pounds) of *CCN-51* was not to our taste but we persevered with a bag of Ecuadorian beans bought in Toronto. That time the chocolate turned out quite good but the flavors seemed too classical or expected, whereas we wanted to produce something exceptional. Then Elfi had the opportunity of attending the newly created *Salon del Cacao y Chocolate* in Lima, where she met many Peruvian cacao growers as well as budding chocolate makers. Because of Peru's renowned cacao varieties and our familial connections, we decided to focus on the Peruvian Amazon region to source our beans and started making a bar using cacao from Ayacucho, and then from Piura. We formally created the company in February 2017. We gave it the name *Qantu* because it is a symbol of unity and hospitality in the Amazon region and it is also the national flower of Peru and Bolivia.

We quickly sent samples to the *Academy of Chocolate Awards (ACA)* in London so that we could have some objective assessment of our production. We were overjoyed to receive two awards on this first try. This immediately propelled us to the forefront of the bean to bar category, and we received orders from here in Canada, but also from the US, France and other EU markets.

HOW WOULD YOU DESCRIBE YOUR BUSINESS TODAY?

MAXIME S. - After barely two years in operation we are producing nearly four tons of chocolate on a yearly basis and we have continued to win awards. On the financial side, we are well past break-even point and looking forward to expanding in new premises, hopefully within the next twelve months. We have one full time employee and take on part-timers for peak seasons, mostly for the packaging. *Qantu* is our baby and we are looking at ways of improving it. Automating some tasks is on the horizon but we must be cautious.

There are many bean to bar businesses these days and most of us use the same set of machines, the stone grinders from the US maker *CocoaTown* or the Indian company *Santha*, a continuous tempering machine from Italy (*Selmi* or *FBM*), and a self-made winnower. What sets us apart and makes us really different is our unique connection and link with farmers in Peru. We do not buy through a broker or importer. We first visit the farms, we check the way farmers live, how they take care of their trees, etc. We sometimes advise on fermentation and in the end we verify every bag–yes, every bag. We pay attention to every little factor, for example, we try to avoid buying right after the rainy season because we've noticed beans tend to have a higher moisture ratio, etc.... With our increased production level we are now capable of buying a small container at a time. Shipping is fast as we usually receive it in Montréal in less than one month. We still need to check every bag and slightly adapt our process according to its characteristics.

QANTU MOLDS

WHAT ARE YOUR MAIN ISSUES / OBSTACLES?

MAXIME S. - We have limited premises and machinery, so at times production can become a bottleneck. This is why we're looking for a new location. At our level of quality, it takes time and a lot of attention to produce chocolate, so keeping up with demand is our main focus.

So far, sourcing has not been an issue, because as I mentioned earlier, we have unique on-the-ground connections which allow us to obtain exceptional varieties that create unmatched cacao sensations. With our current way of buying directly, we are confident that we can continue obtaining amazing cacao beans. This direct approach also guarantees that the five to six US dollars per kilogram (2.1 pounds) that we pay, do end up in the farmers' hands and are not lost along the way to some intermediaries. This represents another guarantee of our ethical commitment towards farmers.

Our goal is to bring more high quality, unique chocolate products to the market. I believe there is a lot of unsatisfied demand for good chocolate, even at prices of eight or even twelve Canadian dollars per bar.

USING A "GUILLOTINE FOR A CUT-TEST TO CHECK QUALITY

CERTIFICATION AND AWARDS?

MAXIME S. - We only use organically farmed cacao beans and we know it because we've seen the trees, the fields, and the people who grow them. This is a kind of self-certification, if you will. But we are not focused on the certification stamp. It is far too expensive for most of the small farmers we deal with. Besides, they do not like chemicals either because some of them live virtually among the trees.

As for *Fairtrade*, it is not applicable to us. It seems to help large retailers rather than small farmers. In Peru, the problem of the young generation leaving the cacao fields to make a daily revenue of one hundred soles (approximately $30) as taxi drivers in Lima instead of thirty-five soles (about $12) at the farm, will not be solved with a ten percent increase in farm salaries. Even when the *Fairtrade* certification really results in a ten percent increase in the price paid to farmers, this is negligible. What we are paying is twice and sometimes thrice what large corporations pay. This has an impact.

Concerning the *Chocolate Awards*, we would be unfair not to value them. I do not think we would have had the swift start we've experienced without the accolade from the *Academy of Chocolate Awards (ACA)* we obtained right at launch time. We continued to participate and have continued to be recognized, including with the *Golden Bean Award* of the *ACA* in 2018. We appreciate the process and the effort because it is a kind of qualification test for us. Having said that, we are not obsessed with medals and awards. We would not create a bar just to please a panel.

RECEIVING ACADEMY OF CHOCOLATE AWARD

MONEY

MAXIME S. - As stated earlier, we are way past break-even point, and we are self-financing our expansion into new facilities and buying equipment.

OTHER TOPICS

MAXIME S. - Although I believe that the current growth in bean to bar or quality chocolate is set to continue for quite many years, I think more of the return must go back to the farmers. The growing number of people ready to pay ten or more Canadian dollars for a chocolate bar clearly shows that there is a demand for quality rather than quantity. This trend seems ready to continue for a while. But if the families and communities where the quality beans are grown do not drastically increase the return they currently receive for their work, there will be nobody to replace the aging farmers soon going on retirement, and there will be no cacao beans in ten to fifteen years. Our close and direct link with farming communities on the ground in Peru provides this clear picture. The change has to come quickly and from all the makers. Say there are about one thousand craft chocolate makers in the world and each buy about ten tons per year; with ten thousand tons we could keep quite a few cacao farmers happy to stay in their community.

FOUND THE PERFECT POD!!

MONICA LILIANA, CARLOTA CHOCOLATE – SANTANDER, COLOMBIA

CONTEXT

IN THE EARLY 2010's, as Colombia's political landscape was starting to move towards a national agreement that would bring peace to the whole country after three decades of internal instability, a wave of optimism was sweeping the country.

Monica Liliana Gómez was running *Mariana Cocoa Export* and started *Carlota Chocolat* with the help of her three sisters, turning their team into the "three musketeers of Colombian chocolate"!

Now exporting to the main chocolate markets of the world, *Carlota Chocolat* showcases the best Colombian cacao beans transformed into award winning chocolate bars.

HOW DID YOU START IN CHOCOLATE?

MONICA L. - In 2013, my three sisters and I were already heavily engaged in cacao sourcing and promoting cacao of Colombian origin abroad, within the framework of our cacao export business, *Mariana Cacao Export*. We decided to follow the value chain and started our own chocolate company to capitalize on the work we were doing with the beans.

We choose our niece's name, Carlota, to remind us of family and to keep in mind the fact that we are working for the future generations. *Carlota Chocolat* produced its first commercial chocolate bar in the Colombian province of Santander in 2013. We acquired a grinder form Santha, and then a larger one from *Cocoatown* as well as a tempering machine and other equipment and started on a small scale but always with quality in mind. The market response was immediately positive and that helped us move forward.

THE FOUNDING TEAM

HOW WOULD YOU DESCRIBE YOUR BUSINESS TODAY?

MONICA L. - I believe we have built our brand as a producer of aromatic chocolate. Like the export business, Carlota is centered on quality, and for chocolate this means aromas and varieties. So we produce four regional origins from four Colombian provinces, and export the chocolate bars to Europe, Canada and beyond.

We now have a small but expert team in Bucaramanga entirely devoted to chocolate production—which averages about a ton per month. They are capable of exploiting the essence and aromas out of the many different cacao beans that we continually discover in various regions of the country. The connection between *Mariana Cocoa Export* and *Carlota Chocolat* allows us to source the best beans available and to help farmers improve their post-harvest processes in order to get the best out of their production. This is a very beneficial collaboration that is unique, at least in our part of the world. Training the cacao community in general, is part of our goal.

TRYING AND TESTING

WHAT IS YOUR APPROACH TO CERTIFICATION AND AWARDS?

MONICA L. - Certifications are a necessity for us because we are addressing export markets that would not even talk to us if we weren't certified *Organic*, Fairtrade among others. We do buy from certified growers. Moreover, if we like a crop from a farmer who is not certified, we will help him get the designation in order to do business with him.

Awards are another topic. Coming from Colombia, which is a relatively unknown cacao origin, it is important for us to be recognized in the countries we want sell to. The Award certificate validates our declaration of quality and it gives potential customers assurances that our chocolate is at an international level. Over time, we have obtained several *International Chocolate Awards*.

From a personal point of view, the award process pushes us to be at our best and incentivizes us to apply a strict discipline in our production process.

CARLOTA'S CURRENT ORIGINS

MONEY

MONICA L. - At its current production level, *Carlota Chocolat* is self-sufficient; but we are always planning for growth and to increase production by adding more cacao origins from within Colombia. The close relationship we have with the cacao export activity of *Mariana Cocoa* is wonderful to help us find new sourcing areas, new farmers and in general because the two businesses feed each other.

OTHER TOPICS

MONICA L. - We see a bright future for Colombian cacao as well as for *Carlota Chocolat*. Colombia has entered a new phase in its political development that is already bearing fruit, even though there are hiccups along the way. Thanks to a stabilized environment, farmers can now access and work in areas that were out of reach until only a few years ago.

Climatological and geological conditions in many parts of Colombia are ideal for cacao farming. On top of these positive factors, Colombia is a relatively little known cacao origin on the international scene so it will benefit from the discovery and rarity effect when expanding worldwide. So, I can see in a not-too-distant future *Carlota Chocolat* becoming a leading chocolate maker in South America in the same way that *Pacari* or *República del Cacao* are in Ecuador.

PROMOTING NEVER STOPS!

ANTOINE MASCHI, CHOCOLAT ENCUENTRO – PARIS, FRANCE

CONTEXT

BASED ON THE outskirts of Paris, *Chocolat Encuentro* operates in one of the oldest chocolate markets in the world. Beyond its pride and reputation for high cuisine, and finger-licking pastries, France has a long history in chocolate. According to the French, the fact that their consumption per person is merely two-thirds that of the top eaters (UK, Ireland, Germany and Switzerland) is a proof that they have a better appreciation of quality over quantity, not a lack of love for chocolate. After all, it was the French who invented the ganache! What else is there to prove, chocolate wise?

The French taste for chocolate skews towards dark chocolate and strong cocoa sensations, not flavored sweet bars. Indeed, France has the highest percentage of dark chocolate consumption in the world.

Like the French mass market chocolate industry, the quality and aromatic chocolate segment is dominated by a handful of long established and renowned companies, like the oldest single origin maker *Bonnat*, the crowned *chocolatier François Pralus* or the *Maison du Chocolat* founded in 1977.

Making chocolate without having graduated at a proper professional school or interned with a Master *chocolatier* is regarded suspiciously. In France, the bean to bar concept of not respecting the rules and tweaking production is only tolerated if the end result is really outstanding. Chocolate is a serious pleasure not to be handled carelessly.

HOW DID YOU START IN CHOCOLATE?

ANTOINE M. - Like most people, we've always liked chocolate. In 2008, during her last year of engineering school in France, my wife Candice decided to take a one-year internship in the Dominican Republic. I quickly joined her. We took advantage of this expatriation to travel around this magnificent island and ended up discovering the cacao fruit. We were fascinated. Back in France a year later, we knew we wanted to go back and start something there. Finally, during the summer 2012, we both left our jobs and flew back to the Dominican Republic to build a chocolate factory with two friends. This was the beginning of the so-called bean to bar movement, and there wasn't as much information available as there is today. Based in Punta Cana, on the eastern side of Hispaniola, we were working with organic local cooperatives to make our chocolate. We were welcoming up to four hundred visitors a day in our factory. We were learning on the job, and it was exhilarating! However, a year after we launched, we were forced to abandon our factory because of local mafia crime, and we went back to Paris. We have not had any more contact with our factory and employees since then. Living in Paris, it took us four years to finally decide to jump again in the chocolate venture. Now with two children, we wanted to create something we would be proud of. This is how *Chocolat Encuentro* was born in 2017. We wanted to start again where our venture had stopped and share our passion for the cacao fruit and for chocolate. This is why we chose the name *Encuentro*. Probably the best word to explain our approach and our history.

HOW WOULD YOU DESCRIBE YOUR BUSINESS TODAY?

ANTOINE M. - Located in a close suburb of Paris, we are still very young, merely two years old, but we have managed to create our personality in terms of chocolate. We currently produce only two-ingredient chocolate, nothing else added. No added cacao butter or emulsifier. We want to be as close as possible to the bean. But it is not as easy to make chocolate as you might think because we must adapt our process to the characteristic of each cacao bag we receive. Also, because the natural amount of fat varies slightly from a crop or an origin to another, we must adapt our process and settings for each new batch. We purchase beans from six countries - the Dominican Republic, Haiti, Guatemala, Madagascar, Bolivia and Réunion Island. We are looking for beans that have personality and aromas we like. Our goal is not to cover every cacao producing country but to generate mesmerizing sensations.

Our objective is to create a truly sustainable and ethical business based on producing really aromatic chocolate. It is important to us that we meet and know the people who grow the cacao beans we use. We know they are produced in an organic and sustainable way. For example, our Haitian beans are produced by a local and relatively new company called Pisa, who pays the farmers about three times the price they used to receive before this fermenting center was created. Because of our relatively small quantities, we often buy via specialized brokers whose ethics and goals match ours. They apply the same strict qualitative criteria we require.

THE" MELANGEUR", THE HEART OF OUR OPERATIONS

WHAT ARE YOUR MAIN ISSUES / OBSTACLES?

ANTOINE M. - As I just said, we find sourcing the beans is not our main issue because we are now contacted by cacao bean growers trying to sell their crop. Sourcing is probably the most interesting part of our job as we get to meet amazing people. Thanks to shipping companies, we can receive samples in Paris from great producers in three days! The hard part is then testing and selecting the beans.

To discover new chocolate sensations is nearly intoxicating! But this requires producing five to ten micro-batches in order to find the ideal roasting and conching profiles. We produce really different and qualitative chocolates, and the demand in France for that kind of product is currently growing.

So far, we have found the commercial side of the business challenging, as we are starting from scratch and without a formal background in chocolate. We do benefit from the vast knowledge our illustrious predecessors have spread, and the large and growing educated customer base they have created.

Because we have to adapt the process for nearly each bag due to the specificities of the beans we receive, the production side is sometimes tricky and exhausting. We generally need to make several iterations before we find the right process and the correct settings.

PARTICIPATING IN THE HARVEST

CERTIFICATION AND AWARDS?

ANTOINE M. - One hundred percent of our products are organic. We are only making chocolate from wild or organic certified cacao beans. Wild beans are naturally organic. We are organic by conviction: a conviction that we must eat healthier and that, if we do not go towards organic food, we leave the door wide open to many compromises. It is true that a large proportion of the quality cacao or fine cacao is organic although not certified for one of the two following reasons: (1)the high cost of the certification process, or (2) chemical compost or pesticides being fortunately unaffordable by the small producers. We should push farmers to certify rather than use chemicals once they have the resources.

We believe it is our duty to show the farmer that there is a market for healthier/organic and natural products that pay a fair price for these characteristics.

Concerning awards, we have a few from the *International Chocolate Awards* and I believe that they do a good job. The main reason is that being small, our production needs some kind of seal of approval, something that confirms our high level of quality and hence justifies our price difference. Secondly, the French market is very structured and the local culture rewards institutions and symbols. Many of the well-known *chocolatiers* are trained pastry chefs and some have received the well-respected prize of Best Craftsman of France (*Meilleur Ouvrier de France*), a yearly event where craftsmen from all over the country compete in a variety of categories. In this context awards are important for us—who do not have those sanctifying attributes and background.

MONEY

ANTOINE M. - We want to bring transparency to the still relatively unknown world of chocolate making. We started by investing in some high quality equipment so that we could get the best from the cacao beans.

We are producing exclusively small batches distributed in a handful of high quality outlets in France. We are focused on quality and will never compromise to increase volumes. Our *secret* is that we take time, sometime a lot of time, to produce the best chocolate.

The French food culture and the chocolate market required us to position our products and our brand on the highest quality possible.

OTHER TOPICS

ANTOINE M. - *Chocolat Encuentro* exists to satisfy us as responsible human beings. We are seeking a fulfilling life - - not a cash-cow. We select bean origins because we like their aromatic notes and the people behind them. We thrive in creating a fantastic chocolate sensation from them. Being true to an ethic of global sustainability, with the humans at its core, is essential for us.

MR. HO, IMMAIM – CHIANG MAI, THAILAND

CONTEXT

THAILAND'S MAIN AGRICULTURAL resources are the Hevea trees that produce the latex used for rubber goods. However, some farmers have been growing cacao trees since the early twentieth century. Cacao production peaked at fifteen hundred ton in 2001 [23] and declined steadily until 2013; it is currently making a comeback as the government publicly declared that cacao would now be a priority for the country. However, harvest is still relatively small—estimated at less than one thousand tons in 2017.

The main cacao producing regions today are in the south near Malaysia and south east of Bangkok toward Vietnam. In the northern hilly region of Chiang Mai, some farmers are expanding cacao production. Because the weather is more extreme - more dry and wet periods - some specific hybrid beans have been developed and used, such as the *Chumpurn* and more recently the *I.M.1* from the *MarkRin* laboratory.

The enormous number of wealthy tourists visiting Thailand has triggered interest for locally made Thai quality chocolate. Mr. Ho is one of the first craft chocolate makers in the country and is trying to create an association of craft makers to unite the fifty or so chocolate producers of varying sizes.

---

23 *http://www.factfish.com/statistic-country/thailand/cocoa+beans,+production+quantity*

HOW DID YOU START IN CHOCOLATE?

MR. HO - In 2017 I had been running my vegan restaurant for sometime using local produce and local Arabica coffee. Many tourists would ask why I was not making or at least offering craft chocolate since there are cacao trees around the city. One day, a visitor gave me the book written by British chocolate expert Don Ramsey [24]. I read it cover-to-cover and immediately started researching how to make my own chocolate. I spent hours on Google and YouTube, and I learned a lot from all kind of craft makers and bloggers. Bean varieties, equipment, processes and even post-harvest treatment, I absorbed everything.

AIMMIKA'S SYMBOL
FROM POD TO CHOCOLATE

I looked for local beans here in Chiang Mai (*Chumphon* variety), and did my first ever batch roasting the beans in a frying pan, winnowing them by hand and crunching them with my home rolling-pin! I added sugar and the result was surprisingly pleasant

24 *https://domramsey.com/product/chocolate-indulge-your-inner-chocoholic/*

as drinking chocolate. I repeated the process many times and tried to make solid chocolate but without a *melangeur*/grinder and no tempering–the texture wasn't smooth and the result wasn't commercially viable.

In mid-2017, I bought the *melangeur* Spectra 11 with variable speed form the Indian manufacturer, *Santha*. Things changed drastically for the better. However, despite the many, many hours spent trying, tempering on my marble slab remained unreliable and hazardous. A Danish tourist staying a few days in Chiang Mai saw me unsuccessfully attempting to temper my chocolate in the back of my vegan restaurant. He came forward and taught me how to temper using the seeding method (*Editor's Note: consists of cooling down chocolate to the right tempering level by incorporating solid chocolate already tempered*). That was the missing link.

In July 2017, I bought molds and professional packaging for my chocolate bars. At that stage, I had a product I could sell, although I could not find the foil I wanted. I called the chocolate, like the vegetarian restaurant, *ImmIam*, based on my daughter's name meaning *Happiness* in Thai [25]. Initially I only sold chocolate bars in my shop/restaurant. Comforted by the interest and praise from customers, I started selling in retail shops in Chiang Mai in December 2017. I produced–and still do produce–in three cacao percentages: 72 percent, 80 percent and 90 percent - and various origins.

THE VEGETARIAN RESTAURANT AND CHOCOLATE LABORATORY IN CHIANG MAI

25 *https://www.facebook.com/ImmAimVegetarianCafe/*

HOW WOULD YOU DESCRIBE YOUR BUSINESS TODAY?

MR. HO - Chronologically, I was the second bean to bar chocolate maker in Chiang Mai after *Siamaya Chocolate*. I now produce about eight hundred chocolate bars monthly. I have recruited an employee exclusively for the chocolate business within the restaurant.

I want to further expand, and for that purpose, I have created with some friends the *Organic Chocolate Company Ltd.* For a chocolate production site near Bangkok we are buying a professional winnowing machine from *CocoaTown*, a larger melangeur/ grinder (Spectra 25) and a tempering machine. Besides increasing volumes we want to get involved at the farmer's level, especially at the fermentation stage–which is crucial to the aromas and the distinctiveness of the chocolate.

As being *organic* is our essence and our paramount goal, we are pushing farmers toward agroforestry. By mixing cacao trees with mango, durian, macadamia nuts, etc., we obtain new and diversified aromas.

CERTIFICATION AND AWARDS?

MR. HO - In Thailand farmers are too poor to buy chemical compost or industrial pesticides. So everything is organic by default. However, the cost of obtaining a certification prevents cacao farmers from going through the certification process. So we do not use any organic certification, but our chocolate is organic. Because of the small size of the bean producers and thanks to Thai laws, there are no child labor or slave labor issues here. Also, the cacao farming sector is too small to be identified as having financial problems, so "Fairtrade" would not make sense at this time.

Concerning awards, I think they are a good thing. In my case, it is far too soon, I am not sure I am yet up to par to participate. I need to improve further before competing. But it would be beneficial as it is a strong motivation to improve. Maybe one day.

MONEY

MR. HO - My chocolate business is standing on its two feet by now. I mean, it is profitable. But it is not my main revenue making activity because the restaurant remains a good venture. However, the ratio is moving all the time. If the *Organic Chocolate Company Ltd.* grows as expected, things may change in the future.

OTHER TOPICS

MR. HO - Now that the Thai government has declared cacao as an agricultural priority, things should move in the right direction. But I hope things will change because I feel currently officials seem more interested in courting large foreign firms than working to help the small land owners. Thailand will progressively be recognized as a cacao country and, maybe one day, be a member of the ICCO!

Another bean to bar producer, in Bangkok - *Kad Kokoa*, is starting an association of Thai chocolate makers. On the Thai chocolate makers' Facebook Group there are about fifty members. The goal is to share technical knowledge that is in high demand in Thailand right now.

NICHOLAS ST. CLAIRE DAVIS, ONE ONE CACAO – JAMAICA

CONTEXT

JAMAICA, THE THIRD largest Caribbean island after Cuba and Hispaniola, has just under three million inhabitants and is visited by about 4.5 million tourists every year. The latest evidence shows cacao was originally introduced to the island by the *Taino* who travelled from South America to the Caribbean twenty-five hundred years ago. It was then planted more extensively on the island by the Spaniards in the seventeenth century. In the nineteenth century, large British-owned cacao companies like *Rowntree* and *Cadbury* owned plantations and chocolate factories on the island. However, Jamaica's most well-known agricultural export is coffee. A law passed in 2018 resulted in the Jamaica National Cacao Board closing and a new regulatory body being formed with a plan to divest the four post-harvest centers previously operated by the JNC. No buyers have been found for the properties and this new organization, called the Cocoa Industry Board, currently works alongside the Ministry of Agriculture to export cacao beans. This is creating uncertainties for the Jamaican cacao industry at a time when it is facing a potentially existential threat from the deadly Frosty Pod disease.

In early 2019, there were six bean to bar chocolate makers on the island, all catering for local supermarkets and especially the tourist market, in resorts and hotels as well as on cruise ships. Based in the north east of the island, Nicholas St Claire Davis, is one of the first to establish his brand *One One Cacao* as an example of what quality crafted Jamaican chocolate can be.

HOW DID YOU START IN CHOCOLATE?

NICHOLAS D. - I moved from London to Jamaica in 2012, continuing my journalism career as the BBC correspondent for the Caribbean. At the same time, I noticed that the British food retailer *Tesco* was selling chocolate bars from the island of Grenada. Being of Jamaican origin myself, I was curious to understand how this smaller territory managed to access the U.K. chocolate market. During an interview with the *Grenada Chocolate Company*, I became very impressed by the way the US-born founder, Green Mott, had been able to work with local farmers and federate their efforts rather than having them compete against each other. As I was congratulating him for this success he challenged me to do the same in my ancestral island of Jamaica.

I am indeed a great chocolate consumer and, at about the same time as the interview, I was hit by a food allergy that prevented me from eating industrially produced chocolate because of the many additives involved. Having cacao beans available in my neighborhood, I decided to make my own one hundred percent additive-free chocolate. Using online resources like *The Chocolate Alchemist* and a pestle and mortar, I started my chocolate journey.

The initial results were good enough for my friends and relatives to push me to go further. Subsequently, I learned more about the techniques and equipment, essentially online, via trial and error, and by visiting many artisan chocolate makers. More than the chocolate making process, I found learning about the bean the most difficult and interesting aspect, together with teaming up with local farmers.

"ONE ONE" COCOA PRODUCTS

HOW WOULD YOU DESCRIBE YOUR BUSINESS TODAY?

NICHOLAS D. - Today, I'm way past the stage of glorified chocolate hobbyist as I produce and sell about three tons of chocolate per year. For me this is roughly two batches per week, but I see more potential. Only four years ago, there was only one chocolate maker on the island, now we are six. There are about seven thousand farmers growing cacao who also want to sell more. Granted, not all of them come up with quality aromatic beans, but still, there is a lot of room for development.

When I started, I did not have a quantitative goal. Beyond being capable of producing good chocolate, my challenge was to be able to work closely with local farmers. Despite having my parents and my roots here, the fact that I relocated to Jamaica after years in the U.K. made me an outsider. After various unsuccessful trials, I now closely work with two estates in the vicinity of where I live, and I enjoy close links with the many professional institutions and structures on the island.

HARVESTING TALL TREES

WHAT ARE YOUR MAIN ISSUES / OBSTACLES?

NICHOLAS D. - I am not really limited in production by my equipment because I could invest in new ones and also hire local employees. The standard salary here is about US$18 per day. Because there is currently a lot of interest in craft quality chocolate from the tourists, I cannot say marketing and selling is a challenge.

My recurring issue is sourcing good beans. There are a lot of cacao beans in Jamaica. Many of them are of good variety and could be wonderful. But many become poor chocolate because of bad fermentation and subsequent processes. Since the secret of quality craft chocolate is more in the bean than in the skills of the chocolatier, everybody is looking for good beans. This creates some competition which, I hope, will result in improved farming. Currently, however, the main issue for farmers is to fight off the deadly Frosty Pod. This could be fatal to many of them.

CERTIFICATION AND AWARDS?

NICHOLAS D. - I do not buy certified beans. In fact, I do not pay much attention or credit to certifications in Jamaica. There are multiple reasons for this. First, I know the farmers I buy from. I know how they live, how they work, etc. I can see for myself that they do not use chemicals as they honestly can't afford the inputs.

As for social sustainability, this is my goal. I want to reenergize the Jamaican cacao sector. I do not negotiate on the price. But I will not negotiate on the quality either. I'm not in chocolate to become rich - that would be a poor choice of activity for that goal - I am in chocolate in Jamaica to bring my little contribution to Jamaican society, to the island I'm coming from. If I were in the US, I would form a certified B Corporation (*Editor's Note: A benefit corporation is a for-profit corporation with a mission that includes positive impact on society. A well-known example is Patagonia.*)

Concerning awards, in 2017, I obtained a Silver and Bronze Award from the Academy of Chocolate in the bean to bar category. It was a way to have my quality recognized, to have my skills recognized. Now I would not pay to enter in a chocolate award competition. I think there are too many of them and I'm not fully convinced by the seriousness or, better said, by their rigor in tasting because it seems like anyone can get a gig as a judge. What I would pay for is an aromatic analysis by a university or a research center that would provide a detailed description of my chocolate. It is not easy to have a good, thorough, and detailed tasting done.

MONEY

NICHOLAS D. - I am still in business and expanding, that said, I've seen people better funded and more skilled fail; so every day is a blessing, sometimes a curse.

OTHER TOPICS

NICHOLAS D. - I would like to come back to sourcing and the farmer's perspective of chocolate. As long as farmers are not recognized as the key element that they are, the chocolate industry and, by extension, society as a whole will not benefit. Even with the current wave of bean to bar and tree to bar businesses from the cacao producing regions, local farmers and local communities are not involved as much as needed. If you look at most of the new chocolate makers in cacao producing regions, they are mostly owned and/or managed by non-locals. *Marou* in Vietnam (*Editor's Note: a US-French team*), *Madecasse* in Madagascar (*Editor's Note: a US founder, now partially owned by US retailer Whole Foods*), *Pisa* in Haiti (*Editor's Note: a French-Haitian*), etc. In order to fully bring all the benefits of cacao farming back to the growers, this should change.

"ONE ONE" CHOCOLATE BARS, STRAIGHT OUT OF MOLDS

BEN RASMUSSEN, POTOMAC CHOCOLATE – WOODBRIDGE, VIRGINIA, USA

CONTEXT

THE AMERICAN CRAFT chocolate market is very active. Although the overall consumption number lags in comparison to most European markets, the statistics hide huge differences. In particular, both coasts have a higher than average chocolate usage. The Beltway (Washington D.C., northern Virginia and southern Maryland), is one of the wealthiest areas in the country, home to many "bean to bar" makers, and enjoys a high level of chocolate spending.

Here, the bean to bar movement is part of a global evolution of food consumption and eating habits toward true sustainability. A growing portion of the public is seeking products that are in line with their social values and ethical goals. At the same time, they are open to new sensations—which allows many entrepreneurial chocolate hobbyists to dive into the bean to bar market.

Potomac Chocolate was among the first wave of such companies and is now enjoying a high level of notoriety and recognition for its many award-winning aromatic chocolate bars.

HOW DID YOU START IN CHOCOLATE?

BEN R. - Coming from the IT world, I was introduced to fine chocolate by my older brother and his wife in Christmas of 2009. They had recently taken part in a Chocolate 101 class at *Caputo's* in Salt Lake City, and put on a small chocolate tasting for our family. We tasted several bars from makers such as *Amano*, *Domori*, *Patric*, and *Amedei*, and I was blown away by just how different chocolate could be based on the origin of the cacao and the maker.

After that tasting, I started to research and learn more about chocolate and starting holding chocolate tastings for friends. This both helped me to improve my palate but also to experience more and different chocolate as I'd include new bars in each tasting. Before one of these tastings, my friend Justin Smith suggested that we attempt to make some chocolate on our own. I thought this was a perfectly ridiculous idea, but within a month (around May 2010), we had purchased some small equipment and some cacao from John Nanci at *Chocolate Alchemy*, which was and is a great resource for learning how to make chocolate at home. For equipment we started with a small *Santha* stone grinder from a crazy person on Craigslist in Scranton, Pennsylvania, and a bean cracker from *Chocolate Alchemy*. We roasted the beans in my home oven, winnowed using a bowl and a hair dryer, and tempered the finished chocolate by hand. By July of that year, Justin and I formed the company *Potomac Chocolate* with the goal of producing and selling exceptional single-origin dark chocolate bars.

POTOMAC CHOCOLATE'S ARTISTIC DESIGN

POTOMAC CHOCOLATE LOGO

We released our first bar, a seventy percent dark chocolate using cacao from Upala, Costa Rica, on November of 2011 at a tasting event (at what was then called *Biagio Fine Chocolate* and is now called *The Chocolate House*). Shortly thereafter, Justin left the company and I have continued as a one-man operation. Over the years, I gradually acquired better and larger equipment to answer the growing demand for my chocolate. This included several small *CocoaTown* grinders which were comparable in size to my original Santha, before upgrading to a *Santha Spectra 40*, and then to a *Spectra 65*. After the 65, I decided to employ a completely different refining and conching process and switched to a Chinese-made universal refiner/conche, which is what I am using at the time of this writing. I have also purchased other types of machines, including *Chocovision* and *Savage Bros*. tempering machines. In addition to the equipment that I have purchased, I have designed, built, or modified several machines for making chocolate. These include several winnowers (I've built four at this point), a convection oven-based roaster with a hacked-in roasting drum, two different conches, a cooling cabinet, and a few smaller items such as vibrating tables. I'm currently working on building a much larger, gas-fired roaster. Good DIY skills come in handy in craft chocolate making because the process requires many different machines, which will inevitably break down and need repair every now and then. As many equipment manufacturers are far away or abroad in Europe, India or China, relying exclusively on support from them could leave you without the ability to produce chocolate for many days--if not weeks or months. Obviously, the cost will also be higher, but the real benefit of building and repairing the machines is the increased understanding it gives you of what exactly the machine is doing and how it affects the quality of the final chocolate.

HOW WOULD YOU DESCRIBE YOUR BUSINESS TODAY?

BEN R. - Today, I generally produce about two full batches of chocolate per week for a yearly total that would be over two tons. This includes several single-origin chocolates, and a variety of inclusion bars. I am currently in the process of looking for an adequate space to build a new and improved chocolate factory. Besides increasing chocolate bar production, I have begun to make a collection of bonbons–which are small, usually ganache-filled chocolates. There is a seemingly endless variety of flavors that can be produced with bonbons which can be produced in a wider variety and with lower barrier to launching a new flavor than is the case with bars. Of course, my bonbons are made with chocolate that I produce. When we initially started *Potomac Chocolate*, we did not have a well-defined goal beyond being able to make my own great tasting chocolate. So, overall I'm pretty happy with the current situation of the business, but, although it is profitable, making chocolate has not yet allowed me to completely give up my position in IT. I'm working on it and believe it is an achievable goal.

POTOMAC CHOCOLATE'S RETAIL PACKAGING

WHAT ARE YOUR MAIN ISSUES / OBSTACLES?

BEN R. - At the start, the main difficulty in sourcing beans was that there was a very limited variety of true fine-flavor cacao available, especially in the small quantities that I, and other new small makers, needed. The main supplier (in the US) was John Nanci at *Chocolate Alchemy*. Over the years, as the demand from a growing market of craft chocolate makers increased, a few other importers began bringing large quantities of cacao into the United States. Chief among these are *Uncommon Cacao* and *Meridian Cacao*. These companies are an integral part of the fine chocolate supply chain, as it is very difficult and expensive to import the relatively small quantities of cacao that small makers need. It is equally difficult for small producers navigating the export regulations of their countries, which are generally geared towards much larger producers and shippers. These organizations are trustworthy and focused on the quality of the cacao and the sustainability of the producers. They are transparent and I am confident that the farmers of my beans receive a price matching the high quality of their product that allows them to take good care of themselves and of their community. As an exception, I also buy direct from one farmer I know in Costa Rica. Currently, my main issues are related to the business side of things. I love the production side of things, developing new bars and bonbons, and continuing to learn how to use my equipment to produce the best flavors and aromas. I wish I had more time to spend on it. There are no immediate-limiting issues there, other than that I simply need more space. I'm satisfied with the equipment I have assembled and have a good idea of how I'll increase capacity when needed. The actual business side of things (marketing, etc.), however, is a bit trickier, and is not something that I am particularly interested in or well qualified for. I am happy with my current retailers and distributor, but growing sales is something that regularly takes a back seat to production. I've had a few, rather unsuccessful sales people over the years and am currently still doing most of this work myself. Having someone else handle that as well as various other business tasks would free me up to focus on production and development.

POTOMAC'S LATEST POUCH

ABOUT CERTIFICATIONS

BEN R. - While all of my chocolate is made with organically-produced ingredients, I do not particularly worry about organic certification. I buy from farmers who grow cacao using organic techniques, but who are generally too small to afford the cost of the certification process. My main concern with sourcing is whether the product is sustainable—both for the farmers growing the cacao and for the environment. If farmers aren't paid enough to live with dignity, it is not sustainable. I have the same feelings towards other certifications, such as *Fairtrade* and *Utz/Rainforest Alliance*.

ABOUT AWARDS

BEN R. - I have a nuanced view of chocolate awards. Of course, I love winning awards for my chocolate. I've won several over the years, including an *Academy of Chocolate Award* for that very first Upala seventy percent bar received just a few months after it was released. I've also received awards more recently from the *Academy of Chocolate Awards*, the *Good Food Awards*, and the *International Chocolate Awards*, which I feel is the most transparent and dedicated to recognizing high-quality chocolate. Winning awards is a great morale boost and can also help in sales, particularly for first time customers. But, there are some inherent limitations to the ability of awards to signal quality to customers. This is for various reasons, including the potential for palate fatigue during the tasting (although some awards put a lot of effort into combating this—notably the *ICAs*). Another issue is that there often is variability between batches of chocolates—especially with smaller and/or newer makers. So, the batch that won an award may be excellent, while a subsequent batch may not be as good. There is also the possibility of a maker selecting only the very best of the chocolate for awards submission, instead of a sample representative of what is available in stores.

2017 ICA RECOGNITIONS

MONEY

BEN R. - *Potomac Chocolate* has been profitable nearly from day one. However, it needs to grow a bit more in order to become my sole activity.

OTHER TOPICS

BEN R. - For my single-origin chocolate bars, I am making exclusively two-ingredient chocolate. But I have started making chocolate that varies from that formula—specifically for inclusion bars and for use in my bonbons. Two-ingredient chocolate has become the norm for new craft chocolate makers, but I think there is space for a more differentiated market. I think new chocolate makers should experiment with different formulations, including additional cacao butter, vanilla, and inclusions to find what they like the best. Assuming that ethically-sourced and high-quality ingredients are used, one way is not inherently better than another. I wish more makers would follow their own path rather than the path of some makers that came before. Different people enjoy different tasting experiences. If we want to widen the number of the people enjoying good craft chocolate, we must also enlarge the scope of good chocolate.

CHOCOLATE COMPANIES

CONSIDERING THE INVESTMENTS and marketing budgets involved in today's chocolate candy manufacturing, the top makers are now huge international corporations with large buying capacities.

Like all large organizations, these companies are subject to the volatility of financial speculations and do not necessarily have a unique national identity. For example, the British icon of industrial chocolate, Cadbury is owned by the American food conglomerate Mondelez, which owns a large portfolio of candy businesses including Milka, Toblerone, and Côte d'Or. In 2017, the largest cacao processor, producing 1.7 million tons of chocolate for professionals, Barry Callebaut was originally a Belgian business. It merged in 1996 with the French company Cacao Barry, and the group is now headquartered in Zurich, Switzerland. Callebaut operates sixty manufacturing plants all over the world.[26]

Top 20 Confection Companies (2018)

RK	NAME	COUNTRY	HQ CITY	PLANTS	EMPLOYEES	SALES (US$)
1	MARS	USA	CHICACO, IL	53	34,000	$18,000
2	FERRERO	ITALY	LUXEMBOURG	23	34,543	$12,390
3	MONDELES INT'L	USA	DEERFIELD, IL	150	80,000	$11,792
4	MEIJI CO. LTD	JAPAN	TOKYO	7	10,673	$9,662
5	HERSHEY CO.	USA	HERSHEY, PA	13	16,910	$7,779
6	NESTLE SA	SWITZERLAND	VEVEY	413	323,000	$6,135
7	LINDT & SPRUNGIL AG	SWITZERLAND	KITCHENBERG	12	14,000	$4,737
8	EZAKI GLICO CO LTD	JAPAN	OSAKA	12	5,488	$3,327
9	HARIBO GMBH	GERMANY	BONN	16	7,000	$3,300
10	PERFETTI VAN MELLE SPA	ITALY	LAINATE	32	17,700	$3,086
11	PLADIS	UK	LONDON	34	26,000	$2,816
12	GENERAL MILLS INC.	USA	MINNEAPOLIS, MN	50	40,000	$2,100
13	KELLOGG'S CO.	USA	BATTLE CREEK, MI	46	33,000	$1,890
14	STORCK	GERMANY	BERLIN	3	6,000	$1,859
15	ORION CORP.	KOREA	SEOUL	16	13,165	$1,718
16	ARCOR	ARGENTINA	BUENOS AIRES	47	22,000	$1,200
17	UNITED CONFECTIONARY MANUFACTURERS	RUSSIA	MOSCOW	19	20,000	$1,169
18	BOURBON CORP.	JAPAN	NIGATA	10	4,900	$1,060
19	LOTTE CORP.	KOREA	SEOUL	22	14,798	$1,055
20	VALEO FOODS GROUP	IRELAND	DUBLIN		2,200	$1,026
				TOTAL	725,377	$95,737

GLOBAL RANKING BASED ON BILLING

Note: The billing numbers shown here cover all the activities of the companies. Most of them produce a wide range of food products that include chocolate or cacao content alongside non-cacao items.

26 *https://www.barry-callebaut.com/en/group/about-us/barry-callebaut-glance*

MAKING CHOCOLATE

SIMPLY PUT, THE process consists of reducing the natural acidity and bitterness of the cacao bean via fermentation at the farming location, then roasting the beans to fully dry them and separate their skin while stressing further some scents, and finally reducing the particle size of the beans and some added sweetener (sugar) to a size pleasurable to the palate (below twenty microns). To improve taste and conservation time, the resulting chocolate must also be tempered in order to present a uniform and long lasting texture.

There is more than one method for each of the steps. Farmers and chocolate makers select the one that best fits their volumes, required quality, and financial resources.

YOUNG FLOWERS AND A NEARLY RIPPENED POD TOGETHER

THE HARVEST

EVEN THOUGH THERE usually are two main harvest periods, cacao trees simultaneously carry flowers and pods at different stages of maturity all year round. Trained harvesters need to identify the ripped pods and carefully cut them cleanly, without leaving an open wound. The depth of the wrinkles on the pod and its color provide good indications of its ripeness. Until the pods are opened, the beans inside remain sterile, allowing unopened pods to be kept up to four days before starting the controlled fermentation process. Once opened, the large amount of chemicals in the mucilage (the pulp) and the seeds themselves, coupled with the relatively high temperature in cacao fields, quickly triggers a natural and uncontrolled fermentation process. This is why when farmers open the pods under the trees and bring "wet beans" to the post harvest centers, the fermentation operation must start within five to six hours of collection.

Depending on the local habits, the topography, and workforce available, pods are either opened under the trees where the empty shells are left to rot, thereby attracting the small aphid midgets that perform the pollination of cacao flowers or brought to a treatment center.

The pod breaking work and extracting of the wet beans with their white pulp must be done carefully so that no bean gets cut open. In large industrial plantations in West Africa and other regions, this operation is often mechanized using ad hoc equipment brought into the fields.[27]

Often, the growers start harvesting and breaking open the pods early in the morning and put the wet beans in plastic bags and buckets by the main trail. After noon, trucks, tractors or donkeys from the post-harvest center bring the crop to the fermentation hangar.

BEANS CUT-OPEN DURING POD OPENING WILL LIKELY ROT, RATHER THAN FERMENT

THE FERMENTATION PROCESS

IT IS NECESSARY to ferment the cacao beans in order to prevent molding, to annihilate the embryo of the seed so that it does not germinate, and to create various chemical reactions that initiate the chocolate aromas.

Chocolate made with unfermented beans has a buttery, or even curdled milk flavor, making it unpleasant. Unfermented beans are used for the extraction of cacao butter which is often deodorized for cosmetic or medical usage.

Fermenting is best achieved using thick wooden boxes covered with banana leaves. Every fermentation cycle leaves some germs or chemical components on the sides of the boxes. Gradually a whitish foam-like cover appears, and the boxes are deemed "seeded." Some of this cover can be transplanted to newly built boxes to speed-up their fermenting ability.

It is important to have a large volume of wet beans per box - between fifty kilograms and 1.5 tons - in order to keep the temperature that will naturally climb to up to 122° Fahrenheit (50 degrees Celsius).

The first eighteen to twenty-four hours, an

27 *http://www.iteks.fr/decoupe-cabosse.htm*
https://youtu.be/AX4Wav5iipY

anaerobic fermentation happens, reaching 86-95° Fahrenheit (30-35 Celsius). By monitoring the temperature, it is possible to identify when this stage is concluded, and a warmer aerobic phase starts. To help that second stage, the beans are turned over to incorporate more air while trying not to cool them down—which would stop the whole process. The most common way of achieving this is to place the fermentation boxes in a sort of giant staircase. Starting at the top level, once the time is right, the side of the box is opened and the beans are pushed down in the second box below. After another two days, that box is given the same treatment and its load ends-up in the bottom box where fermentation is completed. Because of possible rain and the need to bolster the temperature, a roof is placed above the fermentation installations. However, the various acids escaping from the boxes must be able to disperse requiring at least one side to be opened.

Depending on the cacao varieties, the aromatic goals, and the weather, the whole process takes five to ten days. When completed, the fermented beans have a moisture content of about forty percent and need to be dried down to a maximum of seven percent humidity content.

DRIED FERMENTED BEANS

FERMENTATION BOXES

DRYING

WEATHER PERMITTING, THE best drying method consists of placing the freshly fermented beans in one-inch layers on nets fixed approximately three feet (one meter) above a solid area under a greenhouse-like structure. The first two to three days it is advised to turn them every hour, and then every six hours. Unadvisable but used practices include drying on the roadside, on open air wooden tables, and above wooden fires. Drying on macadam transmits tar components to the beans, open tables let animals foul the crop, and open fires produce unwanted odors.

By pulling the transparent plastic covering over the drying tables up or down, it is possible to maintain the temperature under 113-122° Fahrenheit (45 to 50 Celsius). Beyond that level, the skin on the beans could dry before the inside, creating a hard shell that would allow the cotyledons to rot.

A MOISTURE METER

In case of persistent rain, weak temperatures, or extra harvest, gas or wood fire dyers are used. Using heat exchangers to avoid odors, dyers gently and slowly roll the beans in large drums where the temperature is brought up to 113° Fahrenheit (40 Celsius).

LARGE DRYING TABLES IN THE D.R.

Washing

A SMALL NUMBER of chocolate makers require their beans to be washed before drying. This process is done either in small round cisterns or in larger pools. The resulting beans look clean and are free of residue. They will lose some of their sugars and aromatic identity, but also most of their acidic acid. So they likely will require a shorter conching time, which also maintains more flavor. However, washing them adds a risk of molding as they are wetter than when they are taken out of the last fermentation box. Overall I am not in favor of washing the beans before drying them because there is no discernible aromatic difference between chocolate made of washed and unwashed crop, and it adds a moisture risk and a manipulation cost–not to mention the ecological issue created by the need for water and the disposal of germ-filled waste water.

Roasting

ROASTING IS THE second crucial operation for the end flavor of the chocolate bar. It is necessary to bring the moisture level of the beans as close as possible to zero, to boost the separation of the skin from the bean–which improves the winnowing step–and of course, to strengthen or reduce some aromas. Every chocolate maker has their little secret about roasting. Depending on the quality and size of the bean, its level of remaining moisture, the machinery involved and the aromatic profile pursued, roasting can last from ten to fifty minutes. As a rule of thumb, the smaller and more aromatic the bean the shorter the roasting time and the lower the temperature. The right roasting profile is the

result of various trial and error attempts. Compared to coffee, cacao roasting is relatively straight forward as it requires lower temperatures (up to 350F - 180 C instead of 420 F-220 C), shorter roasting time (rarely more than 40 minutes) and a less crucial cooling process.

Using a coffee roaster, a simple rotating oven or a classic bakery oven, the cacao beans should be brought as quickly as possible to the adequate temperature because we are roasting, not cooking the beans. Once a chocolatey smell is identified and some beans start "popping" like popcorn, the beans should swiftly be cooled down.

To find the right roasting profile, I move only one parameter at a time, either the roasting time or the temperature. For aromatic varieties, I set the oven on 250° Fahrenheit and place the beans inside. I take some samples out at 10, 15, 20 and 25 minutes, or more if needed, and smell and taste them to identify the best results. If none are satisfactory or I have doubt, I repeat the operation at 300° Fahrenheit, and so on… For large commodity beans, I would start at 340° Fahrenheit and leave them longer in the drum.

WINNOWING

ALTHOUGH USUALLY PERFORMED after breaking the beans into small parts called nibs, this operation consisting in separating the skin from the bean can also be done before. The principle is simple, by blowing some wind or draft on the broken beans with skin, the skin will fly away whereas the heavier nibs will fall. So, all winnowers involve blowing air and filtering the resulting particles. Small artisan winnower can be made by assembling pipes and a vacuum cleaner, and larger ones include vibrating nets to filter the output. Up to thirty percent of the weight of the cacao nibs can be lost in this step. There should not be more than five percent of skin left with the nibs. Considering that the skin is very light compared to the bean, five percent makes an awfully big volume, so I always aim for two percent.

AN ARTISANAL WINNOWER

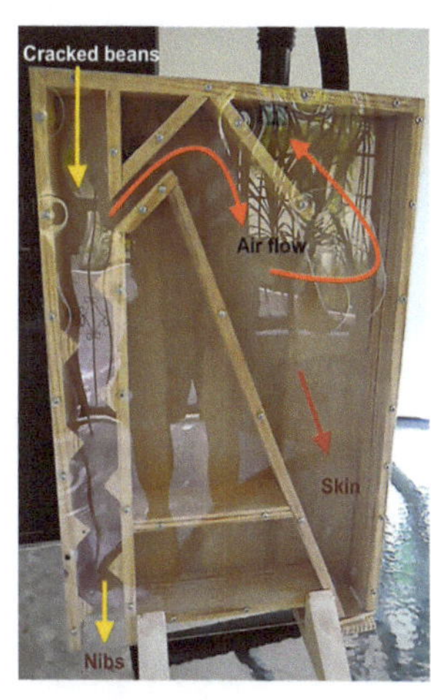

GRINDING/REFINING

THE CACAO BEANS are now clean and broken into smaller pieces called "Nibs". Some chocolate makers throw them slowly directly in their grinder. Most of them though, pre-grind the nibs into a kind of mash called cacao paste that speeds up the next step of reducing the particle size using a refiner or a grinder.

Stone Grinder

A STONE GRINDER is made of two stone wheels turning on a stone bottom is the most traditional way to reduce the particle size. During this exercise the temperature in the grinder climbs naturally up to 113° Fahrenheit (45 degrees Celsius) or more, due to the friction.

Ball Mill

A COMMON OPTION in large operations is a ball mill. It is usually a large round steel tank in which hard metallic balls are placed. The nibs or cacao paste are thrown in the barrel and an internal revolving shaft makes the balls turn, at the same time the temperature is increased via an electric heating system around the cask. By using balls of different sizes and weights it is possible to vary the final particle size at the end of the process.

Ceramic Grinders

ANOTHER TOOL IS the ceramic grinder, similar to a giant coffee grinder. The nibs are pushed through the turning ceramic wheels, which crushes them down. By repeating this process two to three times, a satisfactory particle size below twenty-five microns can be reached.

Each option is better suited to a certain type and volume of chocolate production. The most versatile and most used among aromatic chocolatiers is the stone Grinder.

It is worth noting that some conches (*see below*) are designed to also perform the grinding process using internal scrapers capable of breaking the nibs down to size.

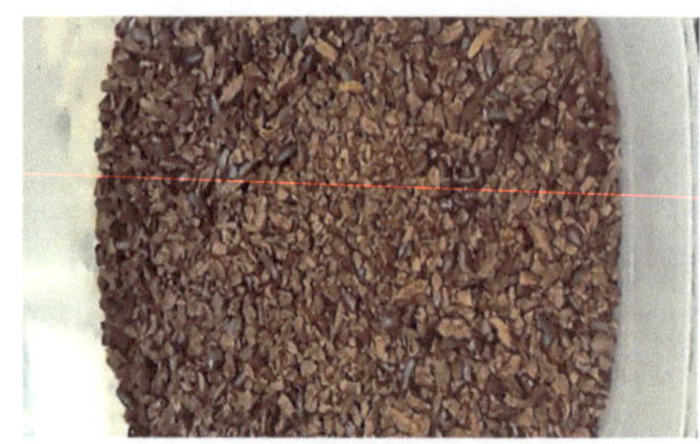

ABOVE, IN ORDER:
CACAO NIBS, READY FOR GRINDER
A VINTAGE LARGE STONE GRINDER
A "SANTHA" 40 POUND GRINDER

BELOW:
A MODERN CONCHE, OPENED

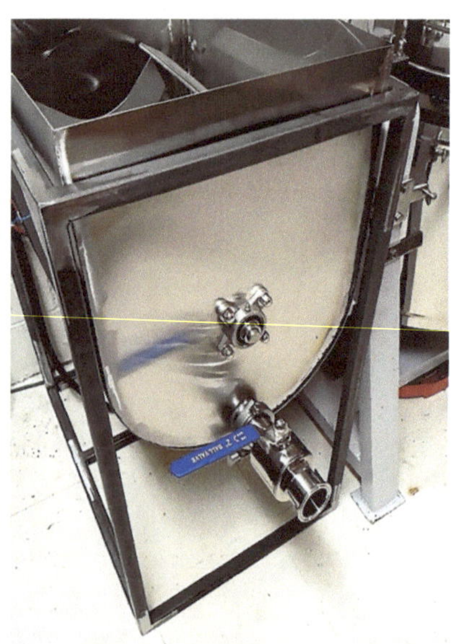

CONCHING

THE TERM COMES from the Spanish word *concha*, for the shell of scallops and other seashell animals. This is because the first Spanish colonizers saw the indigenous people crushing cacao beans using curved trays that resembled large *conchas* and a rollong pin. The word stuck.

Today, the purpose of conching is to mix the cacao with other ingredients such as sugar, and to reduce or

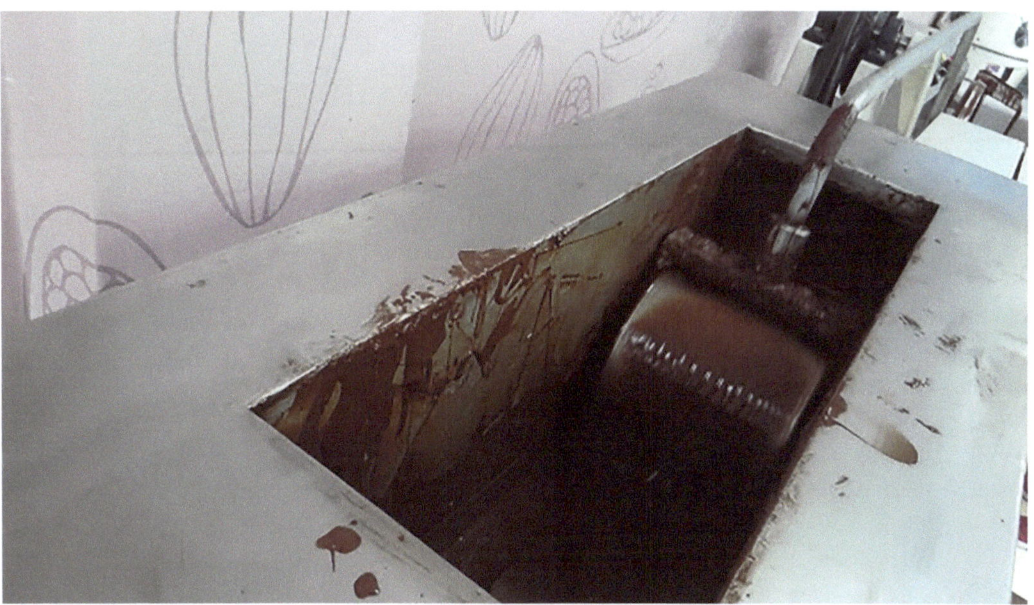

A VINTAGE "MODERN" CONCHE

strengthen some aromatic scents by heating. It is not supposed to further reduce the particle size but rather to thoroughly mix and intertwine the fat cacao molecules with sugar and any other additives, such as vanilla. By heating the *conche* itself, and blowing air on the moving chocolate, chocolatiers are able to evaporate unwelcome aromas and give more presence to others they prefer.

To improve this mechanism, many chocolate makers add an emulsifier. The most widely used is lecithin from some plants. Lecithin is a generic term describing fat substances with the peculiarity of attracting both water and fatty components, making them ideal to create a smooth chocolate texture. Lecithin can be found in animal fat as well as in the skin of many grains and flowers, from wheat and soya to sunflowers. Soya lecithin is the favorite among aromatic chocolate producers who include two to five percent of emulsifier in their chocolate. Note that more than seventy-five percent of the soya currently harvested in the US use Genetically Modified Organism (GMO).

There are various types of *conches*. The first historic ones were a kind of rectangular bucket above a heat source with an agitating roller moving back and forth to mix the chocolate. Nowadays, they are cylindrical tanks covered with a water jacket to monitor the temperature. Inside the barrel, revolving metallic arms move the chocolate while a strong fan forces air in and out.

By adjusting the temperature and the duration of the conching process, chocolate makers can eliminate unwanted flavors–mostly acidity–which allows more nimble cacao sensations to appear.

For industrial makers using larger *conches*, the operation is swift (about four hours); whereas, craft chocolatiers can leave their mix up to ninety-six hours in their *conche*. It is common for them to advertise their conching times as a guarantee of smoothness and quality. However, the double effects of the method do not necessarily generate a great chocolate, unless it was present in the first place.

TEMPERING

COMING OUT OF the conching machine, the chocolate is ready in liquid form. In order to produce chocolate bars that keep for a long time (up to sixteen months under ideal dry and stable conditions), are shiny, and break with a nice snap it is necessary to temper the chocolate before molding it.

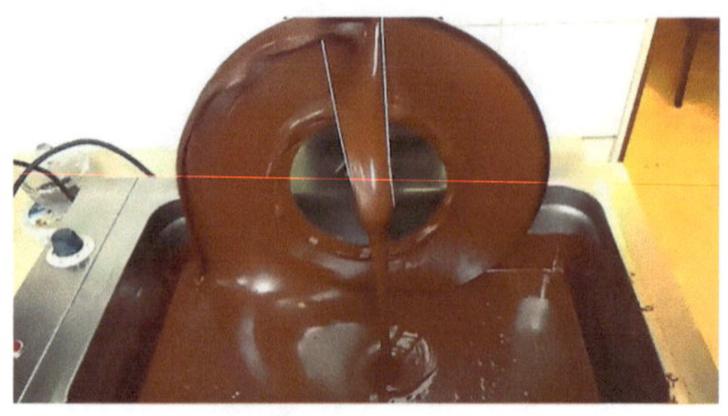

A "BATCH" WHEEL TEMPERING MACHINE

Producing the right crystals

THE FAT MOLECULES in cacao butter have six types of crystallization, or solidification, happening at distinct temperatures. If left to cool down by itself the chocolate will crystalize at different levels, leaving various types of crystals—which results in a powdery and crumbly texture, a dull irregular matte surface and possible white traces of cacao butter.

TYPE	DESCRIPTION	TEMPERATURE
CRYSTAL I	EASY MELT, POWDERY, STRATIFORM	17 C - 63 F
CRYSTAL II	EASY MELT, POWDERY, STRATIFORM	21 C - 70 F
CRYSTAL III	EASY MELT, COMPACT BUT SOFT SNAP	26 C - 87 F
CRYSTAL IV	TOO EASY MELT, THICK, GOOD SNAP	28 C - 82 F
CRYSTAL V	MELTS AT BODY TEMPERATURE, SHINY, SOLID TEXTURE AND PERFECT SNAP.	34 C - 94 F
CRYSTAL VI	TOO SOLID, STURDY. TAKES DAYS TO FORM	36 C - 97 F

The temperatures shown here are approximate because the actual crystallization level varies upon the percentage of cacao butter, and therefore is influenced by the other ingredients in the chocolate (such as sugar, milk, etc.). The standard tempering process consists of bringing the chocolate to about 45 C - 113 F in order to liquefy all the six crystals. Then it is cooled down to 27-28 C - 80-82 F in order to start the creation of Crystal IV and V. By agitating the mix at this stage, the crystals are multiplied over the whole chocolate mix. Finally, the temperature is brought up to about 88° F (31 C) in order to eliminate the initial Crystal IV and retain only the Crystal V.

These variations in temperature can be obtained via many different methods. The basic but most efficient one is to use a heat-absorbing surface, like a marble slab, to cool down the chocolate before reheating it slightly. Another one called *seeding* consists of throwing already tempered chocolate at room temperature into the warm untempered mix in order to bring the temperature down.

Finally, there are many machines providing manual or automatic processes to temper chocolate. The demand by small entrepreneurs for affordable continuous tempering machines has

A CONTINUOUS-CYCLE TEMPERING MACHINE FROM ITALIAN MAKER "POMATI"

resulted in the recent appearance of many such small machines which previously only existed in large format for industrial manufacturers. However, at prices starting at $3,500, they remain exorbitant for the chocolate hobbyist who usually uses more economical batch machines.

MOLDING AND PACKAGING

THERE ARE THREE types of molds for chocolate bars or creations.

Acrylic and silicone molds

THEIR RUBBER-LIKE FLEXIBILITY makes them inadequate for chocolate bars but they are perfect for bonbons and *ganache*. They tend to wear out quickly but are not expensive.

PVC molds

THESE THIN PLASTIC molds, mostly used for confectioneries, are rather inexpensive. Although sometimes used for chocolate bars, they are not very durable and therefore reserved for special occasions like Valentine's Day, Easter, etc.

Polycarbonate molds

THESE MOLDS ARE the professional chocolate bar maker's molds. Sturdy and solid, they can sustain the rigors of professional chocolate making. Some models have magnets that allow then to be handled automatically by conveyor belts.

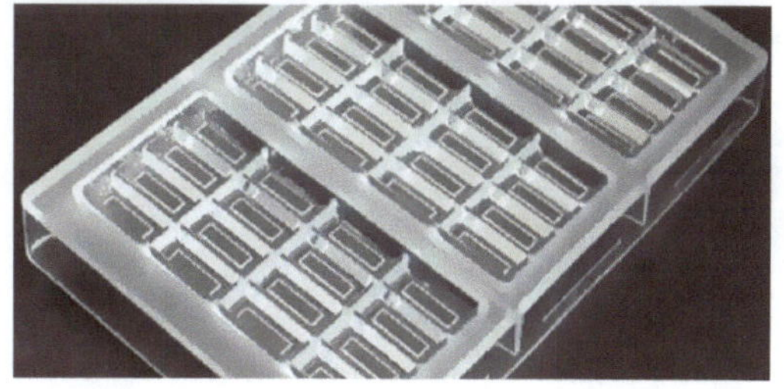

Texture
Commended = C
NAME:
Bronze = B
Silver = S
Gold = G
Super Gold (astonishing) =
COMMENTS

FLAVORS AND AROMAS

BY CHERRIE LO

WHAT IS A GOOD CHOCOLATE?

A GOOD CHOCOLATE is a combination of many characteristics and specificities, and like all human activities includes some subjectivity. For a chocolate to be "good" it should meet the following criteria:

- Look/Appearance: Shows a natural and even shine on the surface
- Sound: Breaks with a crisp and sharp snap sound, confirming a perfect tempering
- Smell: Delivers a rich, multi-layered, complex aroma
- Touch/Texture–(melt/mouth-feel): Melts on the palate smoothly (unless it's a stone-grounded chocolate bar designed to have a sandy texture)
- Flavors/Aftertaste: Lingering pleasant aromatic sensations on the palate, or an evolving and long flavor journey
- No Flaws: No detectable aromatic defects such as cheesy sensations (due to over fermenting), no fungi scents (due to molding), no burnt notes (due to over roasting), etc.

Chocolate profiles

CACAO BEANS HAVE diversified flavor profiles depending on their variety and origin or *terroir*. The aromas of the resulting chocolate are also impacted by the competence and skills of the chocolate maker at every step of the chocolate making process. However, flavor similarities can be found for each major cacao growing country. Here are the common aromatic profiles of the some of the major places of origin:

- Venezuela - Chuao: molasses, dried raisin
- Colombia: sugary, sweet candy, very mild and light
- Peru: maple, butterscotch, prune
- Ecuador: blueberry, blackcurrant, treacle
- Dominican Republic: fig, raisins, nutmeg, cinnamon
- Grenada: muscovado sugar, blackberry
- Papua New Guinea: earthy spices, smoky
- Madagascar: citrusy fruity, earthy
- Ghana: one-dimensional flavor, single tone of chocolatey.

TASTING

TASTING CHOCOLATE IS usually a fun exercise, yet it requires different levels of sensitive judgment and techniques. Having said that, believing in their instinct is the surest way for tasters to really know that piece of chocolate in front of them.

CO-AUTHOR, CHERRIE LO LEADING A TASTING SESSION

Preparation

THERE ARE A few things to take into consideration before tasting:

The night before tasting

- Avoid alcohol, it makes you feel tired the next day and leaves you less focused.
- Avoid a big dinner, a full stomach lowers the quality of rest, hence makes you less energetic the next day.
- Go to bed early and have a good rest. A clearer mind and well-rested body makes you calmer, more sensitive, and aware of your palate sensations. It improves the connection between your palate and your brain, turns your feelings and sensations into a solid judgment and activates the memory associations in your brain.

On the tasting day

- Avoid coffee and any foods with a strong taste, as it affects your palate sensitivity and hijacks your gustative senses.
- Have a proper meal but not a very full stomach–you don't want to feel too hungry during any delicious tasting sessions.
- Be natural. Perfume, nail polish, any scented, after-shave or hand moisturizer will affect your aroma sensations when tasting the piece of chocolate at your fingertips, and it also affects other tasters' aroma detection in the same room.

Environment

- A scent-free environment is the best condition for chocolate tasting. Allow the room to have fresh air flowing in and out.
- The best temperature for storing chocolate is 60° - 68° F or 16° – 20° C. Try to enjoy the chocolate in an ambient temperature not higher than 22° C or 70° F, to avoid chocolate getting soft or melting.

Equipment

Water: Have a bottle of water readily accessible - either still water or sparkling water, depending on your personal preference. Personally, I prefer sparkling water as I think the bubbles carry away any leftover taste from my palate.

Polenta: Polenta is made from grinding corn into flour. It has a rich yellow, yolk-like color, and has a slightly sweet flavor.

Warm polenta paste is an effective palate cleanser popular in the chocolate-tasting world. Its soft, thick and slow running paste-like texture has the magic to rub any leftover taste and carry it away from your palate. Note that texture of the polenta soup is very important, if the polenta is too runny, it won't be able to "rub" your taste buds and carry away your leftover tastes. If it's too solid, you will then need to mash it and it won't do the job effectively. Every little detail counts.

Apples & French Baguette: Though apples and French baguettes have their own subtle aroma and taste, in the award judging session, a lot of time these are used as palate cleansers. It's because the subtle sweetness of an apple slice can clear away and replace any unpleasant chocolate flavors on your palate; while a thin slice of French baguette has the function of neutralizing the acidity and bitterness on the palate, and give you a fresh new start to carry-on tasting.

POLENTA, FLAVOR MAP, SAMPLES

Tasting Notes

WRITING DOWN TASTING notes is important. You will have different experiences throughout the tasting journey; it's always good to record them, compare notes with other tasters, or to compare with your own tasting notes when you try the same chocolate again few days later.

I normally write down my notes in my chocolate tasting notebook. There are also professional chocolate tasting notebooks you can purchase as well.

The five senses of tasting

CHOCOLATE TASTING IS all about sensations—from your eyes, ears, nose, palates and to your brain.

Look: Look thoroughly at the chocolate to see if it has a shiny or matte color, any white traces masking on the outer (sugar blooming), and if there are any air holes. Well-tempered chocolate will give a natural shine to the bar. If there are any white color lines on the surface which look like a mold—it might not necessarily be mold, but more likely sugar bloomed (sugar inside the chocolate got bloom and flowed on top due to poor tempering process and wrong temperature during storage).

Listen: This matters as well for a chocolate bar. Correctly tempered chocolate stored at a preferable "chocolate temperature" between 57° - 65° F (14° and 20° C) will produce a clear snap sound when you break it. Put the chocolate next to your ear and break a piece into two and listen for the sharp snap sound. It's even louder than breaking a biscuit!

Smell: Hold your breath, put the chocolate in your hand. Put your nose into your cupped hands and take a deep breath. The first aroma you inhale is the most organic aroma of the chocolate itself.

Touch: Dark chocolate starts to melt at around 95° F or 35° C (milk chocolate at 82° F or 28° C). Use your finger to touch the chocolate piece and see if there's a greasy feel—low quality chocolate made with added vegetable fat would give you such a lubricious feeling, while craft chocolate (without added fat but only cacao butter) will get your fingers dirty as it melts together with the cacao mass.

Put a small piece of chocolate on your palate and let your body heat melt it naturally. You can use your teeth to break down chocolate into small pieces and then let it melt on your palate, but do not munch and swallow it right away, as you will not be able to taste fully by doing so.

Some chocolates take longer to melt than others due to their fat content and the cacao mass content.

Is it a slow melt? Does it melt smoothly? Is it a buttery, sugary, or a rubbery melt? It's all about the texture and mouthfeel.

Taste: Finally, it comes to the most exciting yet difficult part of tasting, as it usually results in the most diversified answers from different people among these five senses.

Everyone has different sensations; some people are more sensitive to fruit flavors, while others might be more sensitive to natural flavors (woody, earthy, leathery, grassy etc.). There's no absolute right or wrong; the more you taste and train your palate to associate with your brain, the better your sensitivity and accuracy at defining tasting notes. This will lead to more accurate, precise and understandable flavor descriptions. It is important to build a reference system in your brain to link the sensation with a word recognizable by everyone.

TASTE WITH COLOUR® FLAVOUR MAP
BY HAZEL LEE

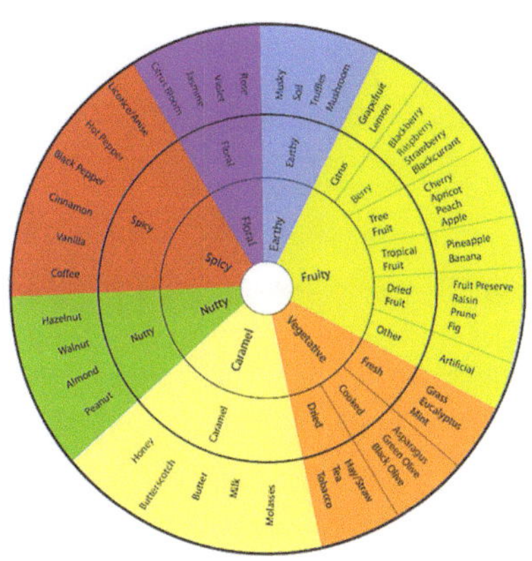

PRINTED WITH PERMISSION
THE C-SPOT® / SENECA KLASSEN
© 2006, 2011
HTTP://WWW.C-SPOT.COM/ATLAS/CHOCOLATE-FLAVOR-PROFILES/

The Tasting Flavor Chart

USING A FLAVOR chart is helpful to guide you in identifying the flavors you've discovered on your palate, and it also trains the connection between your palate sensation and your memories of the flavor in your brain.

There are different tasting flavor charts in the market, from flavor maps divided into color zones to flavor profile colored wheels, to fill-in charts that divide tasting from a variety of categories.

To train your palate professionally and to learn about cacao history, chocolate knowledge and tasting skills, *The Institute of Chocolate & Cacao Tasting* offers one of the most recognized chocolate tasting course across 7 countries including New York, London, Peru and Hong Kong.

The institute was founded by three chocolate leading voices around the world – ICA Award Director Martin Christy from UK, Chocolate/ Tea Taster and Sommelier Monica Meschini from Italy, and the Chef, Restaurateur and Book Author Maricel Presilla from USA.

The IICCT, a U.K. Accredited learning center, the Institute offers three courses in advance levels of «Certificate in Chocolate Tasting». They methodically go through different aspects of cacao & chocolate - from cacao varieties through to chocolate production methods. Throughout the gastronomic sensations training, the courses educate the student's chocolate palate and guide her/him through to gain a real appreciation of the fine chocolate bar tasting.

AWARDS

BY CHERRIE LO

AWARDS CONSTITUTE A great incentive as they recognize excellence in chocolate making and the quality of the components used in crafting fine chocolate. They support companies producing fine chocolate including chocolatiers, chocolate makers, and artisans working with fine chocolate—whether for single origin chocolate bars, flavored bars or bonbons. By helping these high-quality market segments grow and expand, awards also help the farmers that grow fine cacao in the long run.

They represent a celebration of the continuously growing development in the craft and fine chocolate industries and raise awareness among the industry and the public.

There are many competitions all over the world; however, I consider the most reputable chocolate awards in the craft and fine chocolate industry to be the *Academy of Chocolate Awards* and the *International Chocolate Awards.*

AWARDS AROUND THE WORLD

Academy of Chocolate Awards

THE ACADEMY OF Chocolate was founded in 2005 by five of Britain's leading chocolate professionals, united in the belief that enjoying fine chocolate is one of life's greatest pleasures.

Competitions are held in London, U.K., and all entries and competition samples must be sent to London three to four months in advance to undergo a main judging round and Grand Jury judging sessions.

International Chocolate Awards

THE AWARDS WERE founded in 2012 and are run by a group of international partners based in the U.K., Italy and the US, with years of experience tasting and evaluating chocolate, and running fine chocolate-related events.

Competitions are held in a growing number of countries and regions around the world, including Belgium, France, Italy, Germany, the U.K. and the US, as well as Asia-Pacific and Scandinavia. New competitions are added each year.

Winners of the regional competitions are invited to enter the World Final competition where new samples are re-submitted and judged. The World Final designates the best entries of the year.

Both awards are open to any retail products that meet the quality criteria laid down in each competition's rules. Judges include a wide range of chocolate, wine and food experts, chocolate tasters, pastry chefs, food journalists, sommeliers, chefs, and an experienced and professional Grand Jury team.

Once a *chocolatier* has won an award in any of the entries, that chocolate awarded item will be given a certificate that can be proudly displayed in the chocolate factory or retail shop. It also comes with a small shining award logo and sticker that the Award winner is allowed to use to promote the awarded product. The maker must guarantee that the quality and recipe of the of future retail products will remain the same as the sample submitted for competition, as well as provide a time period for the use of the sticker.

This explains why you will see a variety of different shiny stickers on chocolate products in shops. For a beginner who wishes to get to know more about craft and fine chocolates, these are a great tool for selecting among thousands of choices, and act as guidelines to try the best-of-the-best in the world as recognized by reputable chocolate experts.

The Award is especially valuable for small craft chocolate makers and fine *chocolatiers*, as recognition brings visibility, publicity, brand awareness, and ultimately sales and revenue. This is essential for many small, emerging brands to grow and encourages them to pay a premium price for quality beans which, in turn, helps cacao farmers and their communities.

An award for a consumer product also brings visible benefits to the cacao country and its cacao farmers. Here is a real life story.

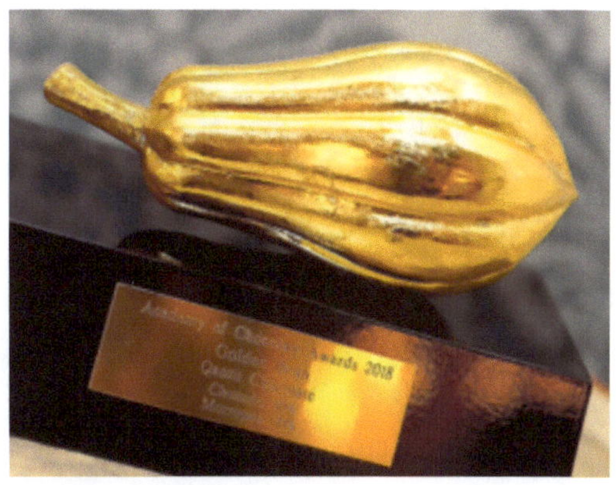

My classmates at the Chocolate School, a Taiwanese couple, Warren and Audrey were Taiwanese Tree-to-Bar chocolate makers. They brought some their own chocolate bars made with cacao grown in Taiwan farms back in the days. None of us had ever heard cacao trees were farmed in such a modern Asian country like Taiwan. With the comments of the course instructor (who happened to be the award director of the International Chocolate Awards) on their chocolate bars, they spent another year modifying their production process and recipe from fermentation, roasting, conching, tempering and molding of the Tree-to-Bar chocolate. A year later, when they submitted their entries for the first time to the International Chocolate Awards, they won eight awards (in 2017). And another ten awards at the Academy of Chocolate Awards the following year. Also in 2018, they obtained the International Rising Star Award in the industry.

Since these awards, which brought them a lot of brand exposure, there is a strong awareness of a new cacao origin, Taiwan. Demand for Taiwan beans has increased dramatically in a very short period of time. At the same time there is a growing amount of Taiwanese Tree-to-Bar/Bean-to-Bar makers in the market. These incentivize local farmers to start growing quality cacao, and shows that the whole award process brings a lot of benefits to the economy at both the business and retail levels.

Nevertheless, some chocolate makers and chocolatiers would like to avoid the game of the award. The well-established brands might not be interested in it as their name, or the chocolate maker/chocolatier behind it, is already famous and they deem their business is strong enough and does not need additional recognition from awards.

Some small companies might not be interested in it as they cannot really afford the competition entry fees, or they do not have the resources and production capacity to take advantage of the potentially huge attention and business coming to them if they win any famous awards.

Types of Entries in Competition

EACH CHOCOLATE AWARD might have slight differences in the categories, but they are usually divided as follows:

- Chocolate Bar (single origin, flavored bar, bar with inclusions)
- Filled Chocolate (bonbon)
- Chocolate Spread
- Hot Chocolate (some include cacao nibs tea)

Judging Criteria

WE JUDGE CHOCOLATE entries from an all-around angle. There are slight differences between each category, though they share the same judging structure, which are very similar to a chocolate tasting. They include:

- Appearance
- Melt and texture
- Taste
- After-taste
- Flaws
- Creativity

For me, tasting craft chocolate bars is like tasting wine, meaning the quality of the main component - the cacao bean - matters the most; whereas tasting bonbons is like tasting a cocktail, beyond the quality of the chocolate itself, there are many other areas to look into.

When judging bonbons, providing comments and grading from one to five, we assess:

- The quality of chocolate (outer shell and filling)
- The quality of the filling
- Whether the shell and filling match harmoniously
- The chocolatier's tempering and decoration skills
- The consistency of the shell, the presence of technical flaws (air bubbles and leaking bottom).

For judging flavored bars and bars with inclusion, we ask ourselves the following questions:

- Is the flavor a really original and innovative concept?
- Does it deliver a fresh and appealing flavor of the ingredient?
- Are all ingredients well-married together and with the chocolate too?
- Do the nutty/fruity inclusions deliver an aromatic roasted/fresh aroma?

THE CULTURAL FLAVOR DIFFERENCES

I'VE SPENT TEN years working in Hong Kong and London in branding and marketing for a wide variety of chocolate makers, from a local celebrity artisan chocolatier to a world-class French patisserie and chocolate brand – Pierre Hermé Paris.

In the course of these years, during my career moves from local to worldwide brands, and my relocation from Asia to Europe, I've seen huge diversity of chocolate types due to the different demands and expectations in distinctive markets.

UK and European Countries

IN WESTERN COUNTRIES, chocolate flavors tend to produce more classic and traditional sensations. The most common and popular flavors, aromas and inclusions in chocolate bars and bonbons are sea-salt, tea infusions, floral scents like rose, violet and lavender, herbs like mint and basil, and inclusions or flavors with roasted nuts like hazelnuts and almonds. This is where the market is today.

Asia/Southeast Asia

INFLUENCED BY ITS extremely varied food cultures, the Asian markets have wildly more diversified and creative flavor sources–which are often challenging to appreciate and accept by western palates.

Each Asian country–like Japan, Korea, Hong Kong, Taiwan, Thailand and India–has its own local savory snacks, with its own range of different national options. So you will not be surprised to discover chocolate creations including savory ingredients like dried seafood and soy sauce. Sometimes culinary dishes, like Green Thai Curry or Coconut Soup, can be transformed or transposed into a chocolate bar.

Savory snacks combining food and nutty flavors (e.g. dried seafood with soy sauce and roasted almonds) are warmly appreciated in Asia and Southeast Asia. This has created a unique style of chocolate with umami taste especially created to serve the demand for sweet/savory snacks that are required by these markets.

THE "WHOLEFRUIT" CHOCOLATE

IN SEPTEMBER 2019, the Swiss headquartered company "Callebaut" launched yet another chocolate product baptized "WholeFruit". This novelty uses all the components of the cacao pod instead of just the beans. Processing the pulp and the juice expelled during fermentation, Callebaut has been able to replace the added refined sugar usually incorporated in chocolate with the natural sugar or sweetener of the pulp. Using this cacao made components allows to elaborate chocolate but also a wide range of chocolatey delicacies like smoothies, desserts, pastries, snacks and more.

Mondelez will soon launch a new range of chocolate products using "WholeFruit", under the name of "CaPao" while Nestlé has already made available the first "WholeFruit" KitKat in Japan.

The selling arguments for Companies using the "WholeFruit" is the absence of refined sugar, which is usually seen as unhealthy, and the fact that nearly 100% of the fruit is exploited which they present as a positive factor. Everyone who's tasted the "WholeFruit" agree that it is very close to existing chocolate and will vary depending on the variety used. As usual, Callebaut does not disclose the origin of the beans nor their variety or varieties and the actual process is not detailed at all.

My take on this new development is that with farmers increasingly disappointed with the low level of revenue they derive from cacao farming, they will be happy if they can get some price hike by selling the whole of the pod instead of just the wet beans. Wether it is better from a farming point of view is debatable. In many plantations the pods are opened under

ABOVE: PHOTO BY WE LIVE CHOCOLATE! ON FOTER.COM / CC BY-NC-SA

the trees and the empty carcasses attract the small insects responsible for the pollination of the cacao flowers. Farmers might need to replace this "magnet" with some kind of sweetener, creating an added workload. Equally, when the pods are opened at the processing center, they are often shredded into some form of organic compost which the farmer will be missing. Regarding the better quality of the sugar coming from the cacao pulp versus that from cane sugar or beetroot, the essential differentiation is in the process to extract the sugar rather than the source itself.

THE NEW CHOCOLATE SCENE

YOU HAVE CERTAINLY noticed that the chocolate shelves in your supermarket now occupy two to five times the space they did a few years ago. At the same time, even though the $1.50, 3.5-ounce (100 gram) chocolate bar still exists, there are many smaller bars selling for anything between three to twelve dollars (some even more). These are the aromatic chocolate bars, some claiming to be single origin or even more exclusive "Single Estate". Also, the dull brown chocolate packaging seems to have completely been banned. Even low-priced bars now come in fancy colorful sleeves. The amount of information of the label has also increased dramatically. Chocolate is not a generic term anymore. The percentage of cacao content has turned into a selling point, and some high-end chocolate makers detail their whole process, up to the conching time, and roasting temperature and duration.

In a process similar to the explosion of microbreweries a generation ago, many entrepreneurial chocolate lovers have turned their passion into business and produce their own delicacies. Thanks to the Internet—which made both technical knowledge and specialized acquaintances easy to obtain—these "new *chocolatiers*" have taken advantage of the diminishing costs of equipment to start producing small volumes of chocolate. Initially *chocolate hobbyists* while they produce for themselves and their friends; they become *craft chocolatiers* or bean to bar makers, when they start selling in the open market, online, on farmers' markets and on retail shops.

They are creating and catering for a new chocolate market segment: the well-informed, health conscious, urban chocolate *connoisseur*. Because of its fast growth, this segment also attracts large and established chocolate manufacturers who are starting to acquire some of the most successful craft *chocolatiers*. For example, the American chocolate giant Hershey bought the Berkeley-based craft maker Scharffen Berger in 2005. Created in 1996 by two cacao passionate Californians, Scharffen Berger is considered the first craft chocolate maker who opened the road to the new chocolate scene.

At the same time, the expanding necessity of cacao growing regions to increase their revenue has led to the multiplication of bean to bar makers working in or near the plantations and addressing local and international markets. For example, the capital of Ecuador, Quito, now counts more than a dozen makers. From the small individual craft maker to the nationally recognized República del Cacao (created in 2005), they all appeared over the last fifteen years.

The online directory of bean to bar chocolate lists over seven hundred small and craft chocolate makers all over the world.[28] In this listing, Africa is currently underrepresented, but the dynamism of the continent is such that I forecast that it will gradually catch up.

More than just the size and quality of their production, what separates these craft chocolate makers is whether they produce from the country or regions where the beans grow and export chocolate or import beans to produce and sell their chocolate products in their country. Both operations are laudable, but the first scenario returns more revenue to the farming country.

CHOCOLATE HOME PRODUCTION

THE PROCESS OF making chocolate I detailed in the pages above can now be performed at home on a small scale, thanks to new offers from some equipment manufacturers and the creativity of some passionate chocolate hobbyists. For each step, there is a scaled-down adapted machine, a tweak using existing kitchenware, or an alternative process. Sourcing quality cacao at a decent price in small quantities used to be difficult but has recently been made less so by the creation of small-scale brokers focusing on quality and sustainability of beans, like Uncommon Cacao[29], Meridian [30], Chocolate Alchemist[31] (small quantities), Cacao Supplies[32] and many more.

Sizing your batch

THE ROASTING AND winnowing steps of chocolate making strips down about twenty-five to thirty percent of the weight of the dried fermented beans; at the same time, twenty-five to thirty percent of a dark chocolate bar is made of added products such as sugar and emulsifier. So as a rule of thumb, if you produce dark chocolate, eight pounds of beans will produce roughly eight pounds of chocolate. If you use an eight-pound grinder/*melangeur* it is a good idea to start with eight pounds of beans.

28 *https://beantobarchocolates.info/cat/chocolate-maker/*
29 *https://www.uncommoncacao.com/*
30 *http://www.meridiancacao.com/*
31 *https://shop.chocolatealchemy.com/collections/cocoa-beans*
32 *http://www.cocoasupply.com/raw-cacao-beans-ecuadorian-nacional-arriba/*

Sorting

SPREAD THE BEANS on a tabletop in a thin layer and throw away anything that is not cacao beans (such as twigs, little rocks, burlap crumbs. etc.). Then, discard the beans that are flat, black, stuck together, unfermented, rotten or otherwise unhealthy. Finally, group the clean and usable beans by size.

EQUIPMENT: TRAYS AND GRILLS.

Roasting

BEANS SHOULD BE grouped by size in order to roast for the adequate length of time. The duration and temperature required varies between 230° - 350° F (110° - 180° C) for ten to forty-five minutes. This will depend on the size and quality of the beans, and the expected flavor profile. For example, a fruity *Nacional* from Ecuador will probably give its best after fifteen minutes at 250° F (120° C) , whereas a large *I-M-1* from Thailand will likely need twenty-five minutes at 340° F (170° C).

EQUIPMENT: A KITCHEN OVEN WORKS FINE BUT SMALL COFFEE ROASTERS ARE BEST. STARTING PRICE $350.

A SMALL COFFEE TOASTER

Breaking the beans

BREAKING THE BEANS down into nibs can be done manually using a rolling pin. With larger volumes it is easier to use a juicer to generate the nibs.

EQUIPMENT: KITCHEN JUICER, LIKE A CHAMPION JUICER. STARTING PRICE $200.

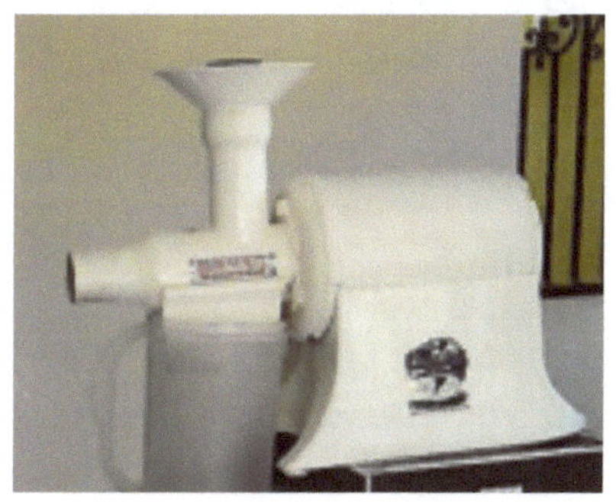

A "CHAMPION" JUICER USED TO CRACK BEANS

Winnowing

TAKING THE SKIN out of small quantities of roasted beans before breaking them into nibs can easily be done manually. In fact, manual winnowing results in less cacao wasted than using a machine. However, this is a time-consuming activity and it is easy to assemble a home winnower system using a vacuum cleaner and some pipes or a wooden box. Equipment: Vacuum cleaner and DIY expertise. Starting price $200.

Pre-grinding

TO TURN THE clean nibs into a coarse paste that will be refined in the stone grinder, a juicer can be used. The friction inside the juicer will naturally bring the temperature of the resulting paste to up to 120° F (50° C). This will help start the particle reduction process in the grinder/*melangeur*.

EQUIPMENT: JUICER, LIKE CHAMPION JUICER. STARING PRICE $200

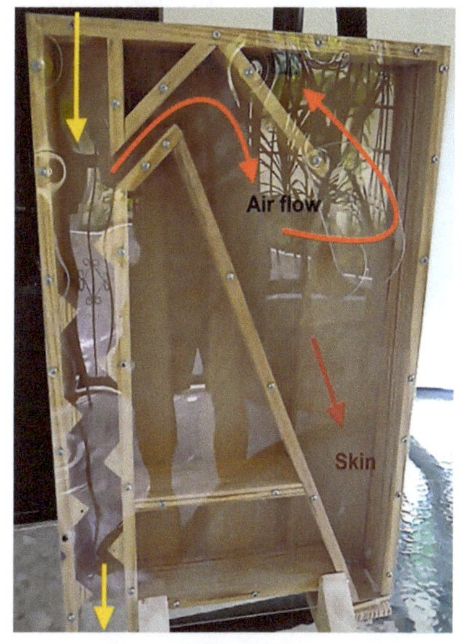

A DIY WINNONER USING A VACUUM CLEANER

Grinding / Conching

THE WARM CACAO paste from the juicer needs to be refined to particle size below twenty microns so that your taste buds do not feel any ruggedness. A small stone grinder is necessary for this step. Although not exactly identical to a real *conche*, these machines can also be used to mix the cacao liquor with the other chocolate ingredients such as sugar and flavors.

EQUIPMENT: SMALL WET GRINDER, EXAMPLE: PREMIER, SANTHA, CACAOTOWN. STARTING PRICE $180.

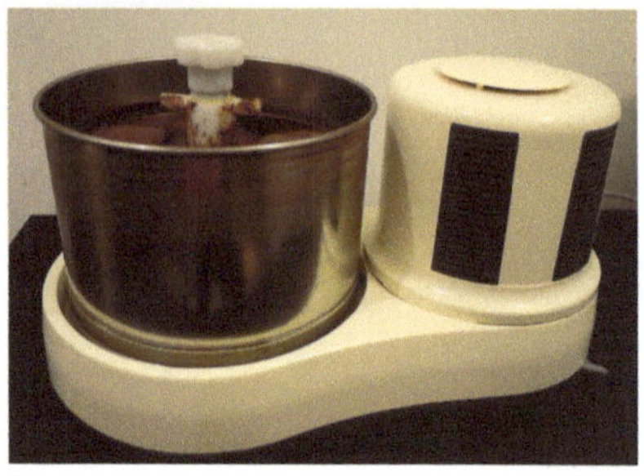

MADE BY SANTHA (INDIA), THE SPECTRA 11 GRINDING 3.7 KG OF CHOCOLATE

Tempering

TEMPERING A SMALL volume of chocolate can be done by warming it over a double boiler before bringing down the temperature by spreading it on a marble slab and finally rewarming the chocolate using the double boiler. As volume increases, it is practical to use a batch tempering machine that will regulate the temperature while the chocolate is constantly mixed. Finally, a continuous tempering machine is the most practical but constitutes a larger investment.

EQUIPMENT: BATCH MACHINES STARTING PRICE $180, CONTINUOUS MACHINES $3,500.

Molding

AS IT IS necessary to shake the filled molds in order to get rid of the air bubble that formed during tempering, the use of a vibrating table is helpful. Professional vibrating tables cost upward of $800, but building one using some springs and an electromagnet as a vibrator does not require a high degree of DIY knowledge. Equipment: Molds starting price $25, vibrating table starting cost, DIY $150.

CACAO 2050

IN A VOLATILE political world stressed by climate change, evolving technologies and financial pressures, forecasting the major trends that will create the cacao and chocolate reality of tomorrow's cacao farming seems like an impossible challenge. Here are our main conclusions:

SUSTAINABILITY AND TECHNOLOGY

IT SEEMS A safe bet to say that by 2050 or soon afterwards, cacao farming will be sustainable or will not be at all. Whatever the problems and issues faced by the large international cacao buyers to implement efficient and successful sustainable cacao farming, I am confident that solutions will have been found and implemented. As there is no silver bullet for such an all-encompassing problem, the means used will certainly cover multiple aspects of farming and involve new technologies.

In this future, there will likely be flavorful or aromatic GMO seeds, irrigation systems monitored via satellite to optimize water usage, improved legal land protection for growers, more involvement from cacao bean buyers into farming and post-harvest treatment, and increases in production of chocolate and chocolate by-products closer to farming locations.

Large companies will need to have a fully transparent supply chain to demonstrate their involvement with farmers, similar to that of the small bean to bar maker reporting that they personally know the person who grew the beans used in the chocolate bar. This will require new information technology in the field, such as blockchain and RFID microchips to operate a reliable but cheap traceability process.

As buyers are more interested in flavors than in varieties, especially in the specialty market segment, there will be a growing marketing value placed on the origin of beans rather than their genetics. This will turn countries of origin and plantations or estates into brands. However, as genetic and agronomic technology become more accessible, farmers will be able to create varieties perfectly suited to their topography, climate, and the organoleptic demands of chocolatiers, hence increasing seed varieties.

THE MOST EXPENSIVE CHOCOLATE BAR SELLING FOR US $250

CHOCOLATE MAKING

Flavor innovation

ALREADY THERE ARE many different ways or tools and machines to produce chocolate. With the advent of accessible small-size equipment, new processes will continue to be invented. There is still a lot to learn at the agricultural and rural levels. For example, pioneers are experimenting with innovative forms of fermentation, and even completely new post-harvest treatments. Although appearing like an oxymoron, a month-long cold fermentation process is being developed with the goal of carrying all the flavonoids and healthy components of the beans to the chocolate bar.

To respond to the demand from new markets such as Asia, new flavors are being created. Because a lot of the chocolate innovations will be technology-driven, their pace is likely to increase with time. This will also permit to manufacture healthier chocolate containing less harming components and more of the good ones, turning chocolate into one of the best foods, namely, medicine. Revisions in food ingredient regulations may be necessary to accommodate these changes. For example, the European Union first banned chocolate makers from advertising their products as health food, but later authorized them to mention that flavonoids improve blood pressure, boost elasticity of blood vessels, etc.

SALES AND MARKETING

THE CURRENT MARKET segmentation of eighty percent bulk beans/chocolate, 19.8 per cent quality beans/chocolate, and 0.2 percent specialty cacao/chocolate is likely to evolve—but only slowly and gradually. It can be expected that in twenty years the specialty and quality segments will represent a slightly higher percentage than today, meaning a global improvement in cacao quality. Considering a yearly increase of ten percent of both segments over the coming years, to the detriment of commodity beans, their volumes would triple by 2040 and lead to segmentation of around seventy percent bulk, 29.5 percent quality, and 0.5 percent specialty, while global production could climb up to eight million tons.

These expected developments assume the price paid to farmers for their beans (the farmgate price) will increase significantly to recover some of the lost income. Depending on the currency used and its associated inflation rate, this farmgate price has been divided by three to six over the last forty years.

Towards custom designed chocolate flavors

LIKE ANY OTHER product, chocolate is more and more susceptible to trends and fashion. Increasingly, to most consumers, eating chocolate represents the pleasure moment of their day, comparable to and often replacing drinking a good wine, and for some, smoking a cigarette. This positioning as a special quality moment to share or to experience daily is being exploited widely by all chocolate manufacturers. To answer that demand, makers are already increasingly offering bite-size, high quality products that are both affordable and aromatically unique.

This expected important growth in demand will create a space for many small and distinctive chocolate brands, specializing in their own flavor and own origin or chocolate type. Some of these chocolatiers will come from the current wave of bean to bar makers because they already meet the demand for locally-produced sustainable and quasi made-to-measure products that the public appreciates.

The Asian Influence

AS A WHOLE, Asia currently represents about one-quarter of the world chocolate consumption. At the current trend, that region will represent well over thirty-five percent of the market in twenty years. In order to grow further in booming markets like China and Japan, new specific aromas and presentations will be developed to meet cultural customs. As a third of the Chinese buyers declare that they prefer foreign brands, there is a huge interest for traditional chocolate makers to tailor their confections to meet this huge opportunity. Incorporating yuzu (reminiscent of citrus), roasted rice tea, flower scents like jasmine and chrysanthemum, and macha powder will spread beyond the Asian market into more traditional ones.

CONSUMPTION

Quality is better than quantity

THERE IS NO reason to forecast a decrease in chocolate consumption. Even if the current big consuming countries, such as the U.K., Germany and Switzerland, do not significantly increase their appetite for chocolate, there is a huge potential for traditional consumers in other European countries and the US to augment consumption.

But the most important growth in consumption will come from relatively new consuming countries: China, India, Russia, and Turkey and from cacao producing countries like Brazil, Colombia, and Vietnam, which are already producing and consuming their locally-farmed cacao products. Seasonality or event-related consumption will increase further. Already sales on Valentine's Day represent twenty-five percent of the yearly total in many countries, including the US and Japan. This will reinforce further the importance of marketing and communications, like the World Chocolate Day.

Despite the expected increase in quantities of chocolate consumed, the average quality will also improve, pushed by farming and production technologies as well as by market demand.

EPILOGUE

IN THE ALARMING global landscape of chocolate, the worst is not certain. It seems that consumers are gradually favoring quality over quantity. The need for farmers to grow better, more aromatic beans, and execute a better post-harvest process is starting to be reachable thanks to the efforts of the small chocolate makers who are capable of paying more, and also to the large buyers who have the resources to bring new technologies to the field. These large businesses are the only ones able to bring all the scientists to the table to research and implement the necessary changes. Genetics, climatology, and automation are all part of the cacao solution. After all, keeping cacao farmers happy and improving harvests represent an existential necessity for the large multinationals.

But these industrialists must better accommodate farmers' interests by dedicating a bigger share of their revenue to them. The price paid to farmers in the fields must go up!

Provided all these stars are aligned, we shall enjoy mesmerizing chocolate forever...

APPENDIX A

FLUFFY CHOCOLATE MOUSSE

Ingredients:

- 7 ounces (200 grams) dark chocolate 65% to 70% cacao
- 1.8 ounces (50 grams) fresh cream
- 3 ½ ounces (100 grams) milk
- 8 eggs
- A pinch of salt

Optional:

- 2 tablespoons of honey or maple syrup
- ½ cup espresso coffee

Tools:

- A large Pyrex bowl, a small pan, a small bowl, an egg-beater.

Process:

START BY MAKING soft ganache:

1. Break the chocolate in pieces and place in a Pyrex bowl. Set microwave oven to 60 percent of its strength.
2. Melt the chocolate in microwave and open to stir approximately every minute. [Note: microwaves vary, it may take several exposures to melt completely; make sure to scrape the sides and stir to mix each time.]
3. Mix the milk and cream and heat in a pan (or in the microwave in a glass container—never heat plastic). Add optional ingredients to this mix.
4. Once the chocolate is fully melted, pour the milk/cream in small increments while mixing with a spatula.
   - This produces a rather liquid ganache.
   - In a large bowl, separate 8 egg whites and beat with a pinch of salt until they become a solid white "snow" (and form stiff peaks).
   - Wait for the ganache to cool down (to about 105° Fahrenheit or below 40° Celsius). Then, incorporate 5 of the 8 yolks by mixing in with the spatula. Discard the remaining 3 yolks.
   - Finally, pour the chocolate mix into the beaten whites and mix gently with a large spatula.
   - Cover the mousse with an air-tight plastic wrap and place in the fridge for at least 4 hours (24 hours is better).
   - For added pleasure, serve with shortbread or yogurt.

CHOCOLATE TARTELETTE

The crust

IN ORDER TO create a pleasant texture and taste, the crust must be crunchy but not too dry. I recommend to create a dough with almonds.

Ingredients:

- Flour: 350 Gr or 12.3 oz
- Almonds: 60Gr or 2.1oz
- Butter: 180gr or 6.4 oz
- Fine sugar: 120gr or 4.2 oz
- 1 egg (preferably organic)
- Some salt

Process

IN A SALAD BOWL mix 1/4 of the flour (90 gr or 3.2 oz) with the butter softened at room temperature, the sugar, the egg and the salt. Once the mix is homogeneous add the rest of the flour. Mix gently. Again, once homogeneous flatten on a marble surface with a rolling pin. About 2 to 4 millimeters or 0.1 inch is the right thickness. Cover with a plastic sheet and let it rest and harden in the fridge for about 15 minutes. Cut it in the required shape. If you do not have a circle of the right size you may have a glass of the right diameter. Place the dough in molds. Unless you use silicone molds, you will need to butter them. Place in the oven at 160C or 320F during 15 minutes or until the crusts reach a toasted color.

While the crusts cook you have time to prepare a soft Ganache to pour inside them. Remember that the Ganache will require at least 12 hours to crystallize.

The Ganache for the filling

THIS TYPE OF Ganache is well suited to use as the filling of chocolate tarts or as the inside of chocolate bonbons. Because it is covered or held by a stronger "envelope" it is softer than the one used for Truffles or bonbons to be dipped in chocolate. The quantities of chocolate vary depending on the proportion of cacao.

Ingredients

- Chocolate – If 70% dark 330gr or 0.7 Pd- If Milk chocolat at 40%, 450G or 1 Pd
- Fresh cream – 300 Gr or 0.7 Pd
- Honey 30Gr or 0.1 Pd or to your taste
- Flavor (Coffee, Thym, Laurel etc..)
- Butter 60Gr or 0.2 Pd

Process:

1. First melt the chocolate to an homogenous "paste". Do not bring it to more than 50 C or 120 F.
2. Then add the flavor to the fresh cream and bring to a boil. Add honey if you're using bitter cacao. Once boiling, pour a quarter of the fresh cream inside the chocolate while mixing gently by drawing circles in the mix. Once well mixed, add an other quarter and so on, until the whole liquid is absorbed.
3. Let it cool down by stirring the mix, and when it is at about 91 F or 33 C add the butter cut in dices. This will create a more unified ganache.
4. Once done, you can the soft Ganache in its "container", be it a tart, a molded "Bonbon" etc...

HOT CHOCOLATE DRINK

A HOT CHOCOLATE drink with creamy texture, strong flavors and a long lasting after-taste is easy to make. It is also easy to adjust to your special taste.

Ingredients:

- 6 to 7 dinner spoons of dark chocolate palets 60% cacao or more
- 3 dinner spoons of whole powdered milk
- 1 dinner spoon of soya flour, or other flour
- 2 dinner spoons of sugar cane
- 1 tea spoon of vanilla extract
- 1/2 pound (250Gr) of milk

The sugar can be advantageously replaced by chestnut honey. Potential flavors include mint, coffee, oregano, pepper etc.

Process

1. Place the chocolate palets and the powdered milk in a mixer and grind until obtaining a granulated mix.
2. Add the sugar cane to the mix.
3. In a pan bring the chocolate mix to fluid and add fresh milk – could be soya milk – to liquify.
4. Bring to a boil and keep it boiling for 1 minute. Serve in small coffee size cups.

www.ingramcontent.com/pod-product-compliance
Lightning Source LLC
LaVergne TN
LVHW070124110826
845147LV00002B/186

* 9 7 8 1 7 3 4 0 0 4 0 2 1 *